Replicate Universe

Science-Based Reality or
Mythical Fantasy?

Replicate Universe

Science-Based Reality or Mythical Fantasy?

Rabindra N. Mohapatra

University of Maryland, USA

World Scientific

NEW JERSEY • LONDON • SINGAPORE • BEIJING • SHANGHAI • HONG KONG • TAIPEI • CHENNAI • TOKYO

Published by

World Scientific Publishing Co. Pte. Ltd.
5 Toh Tuck Link, Singapore 596224
USA office: 27 Warren Street, Suite 401-402, Hackensack, NJ 07601
UK office: 57 Shelton Street, Covent Garden, London WC2H 9HE

Library of Congress Control Number: 2026017873

British Library Cataloguing-in-Publication Data
A catalogue record for this book is available from the British Library.

REPLICATE UNIVERSE
Science-Based Reality or Mythical Fantasy?

ISBN 978-981-98-3254-5 (hardcover)
ISBN 978-981-98-3320-7 (paperback)
ISBN 978-981-98-3255-2 (ebook for institutions)
ISBN 978-981-98-3256-9 (ebook for individuals)

For any available supplementary material, please visit
https://www.worldscientific.com/worldscibooks/10.1142/14861#t=suppl

Desk Editors: Aanand Jayaraman/Gabriel Rawlinson

Typeset by Stallion Press
Email: enquiries@stallionpress.com

In memory of my advisor and mentor
Robert E. Marshak

Preface

Imagination is among the best weapons scientists can have in their arsenal. It gets through to its destination, spurred on by creativity, logic, and mathematics. If observational backing provides a foothold, then the idea can claim itself as an integral part of human knowledge. Two shining examples of this that changed history are the works of Niels Bohr, which connected observations of light spectra in mundane materials to the fundamental inner workings of atoms, and that of Albert Einstein, which expressed the oldest known force in nature as the twisting and turning of planes and lines. Such examples are what built science: Only the luckiest of ideas survive, and science celebrates the steps forward.

Pursuing science solely based on imagination, though, can be risky, and the graveyard of scientific misadventures bears ample testimony to that. It must, however, be said that intuition and creativity always spring from deep inside the human psyche and ought to be encouraged since they are often the well-springs of something new that may have virtue.

This book tells a story, partly based on imagination, whose exploration is in its infancy. It deals with the idea of dark matter in the universe, which is real, but what exactly it is has baffled scientists for almost four decades. This has given rise to speculation about a separate sector of the universe where it may reside. Even though the idea is not implausible, giving dark matter the credit for determining the fate of the entire universe could be among the grandest of reveries. The book takes the view that this alone should not entitle it to be regarded with skepticism. Entertaining new ideas about the universe is not new. Scientists and philosophers have done so over the ages, thinking, among other things, about the possibility of the universe having multiple cousins and the rich diversity they could bring to our experience. Sometimes, proponents have been so deeply committed to such ideas that they have exposed themselves to major personal risks. History records them as heroes and passes their stories on as lessons for posterity.

Assuming the mirror duplicate of our universe as the home of dark matter, though intuitively attractive, could be risky. That is, however, the path we traverse in this book. It is an exploration with no agenda but is merely driven by curiosity to build a stately edifice of two universes together, complementing each other, providing the other what it may lack. The question is: "Is it merely elegant chaos or a first step toward the truth"? The goal is to find some ways that we can use to persuade others about its plausibility. This is clearly a mammoth task, of which only a tiny beginning is being attempted by a small group of scientists. This book is a summary of those attempts.

The book starts with a brief motivating discussion of dark matter and the need for a new framework beyond what we now use to understand nature, namely, the standard model. The attempts to find dark matter as a particle of the visible world, which would extend the standard model, are summarized. The founding assumptions and ideas for a mirror world or a replicate sector to our familiar universe, and why it can legitimately claim to be the home of dark matter, are then laid out. Whatever little is known about the implications and possible tests of this idea is then described, hopefully in a language accessible to non-experts. At present, there is no contradiction between the mirror world idea and observations, but neither is there compelling evidence for it, except possibly for some hints. It is hoped that whatever little indications now exist will eventually multiply. This book, which contains no mathematical equations and is written for non-experts, will hopefully grow the community of believers, one or more of whom can eventually uncover the truth.

If experiments eventually reveal dark matter to be part of the visible universe rather than of the replicate sector, the broader concepts laid out here may still retain value, though they would play a subdominant role in the narrative of dark matter. In that case, they may eventually become a part of the broader colloquy in astronomy. If this book plays even a small role in achieving that, it will be deeply gratifying.

Acknowledgement

This book took shape from my research in the field of dark matter physics and the belief that dark matter may be part of a hidden mirror sector, existing alongside us within the universe we inhabit. I learned most of what I know about the mirror world from discussions with my collaborators, Haipeng An, K.S. Babu, Zurab Berezhiani, Shao-long Chen, Itzhak Goldman, Salah Nasri, Shmuel Nussinov, Nobu Okada, the late Vigdor Teplitz, and Yue Zhang. I thank them for generously sharing their knowledge and insight into the subject, making each collaboration a joyful adventure. My thanks also go to Kaustubh Agashe, Evgeny Akhmedov, Gustavo Branco, Zacharia Chacko, Bhupal Dev, Keith Dienes, Bhaskar Dutta, Anson Hook, Xiangdong Ji, Yuri Kamyshkov, Demos Kazanas, Manfred Lindner, Jogesh Pati, Goran Senjanovic, Robert Shrock, Alexei Smirnov, Raman Sundrum, Hai-Bo Yu, and Yongchao Zhang for discussions in recent years on this and related subjects.

I also thank the World Scientific team, including Steve Anglin, Gabriel Rawlinson, and Aanand Jayaraman, for doing such a professional job of bringing the book to publication.

Finally, I am thankful to my wife, Manju, for her support, advice, and sacrifices during the time this book was in preparation and always.

Contents

Introduction

"You can't cross the sea merely by standing and staring at the water."
— Rabindra Nath Tagore

When we look up at the night sky, we see a panoply of stars — some bright and some faint, some twinkling and some steady, some clustered together, some by themselves — all spread out against the background of a dark abyss, changing their positions night after night. We give them names and track their paths from night to night until we become exhausted and then start again. Stars have been a fascination of humankind for as long as we have been around, and they have been used in the past for more than just satisfying our curiosity about the universe. Polaris, or the North Star, provided a guiding light for the sailors on the dark ocean; the constellation Orion, fully visible in winter, helped farmers to determine the right time to plant crops; Pleiades, also known as the Seven Sisters, was used by priests to identify auspicious moments to perform religious ceremonies. Since then, however, things have progressed far beyond these uses. Today, we study the stars to expand our knowledge of the universe — our cosmic home — and even possibly to warn us of potential astronomical dangers that may be approaching. In doing so, the stars could also offer clues about the future of life on Earth.

The stars that we see are part of what astronomers call *galaxies*, which in turn are often found in clusters of galaxies that populate our universe. Inside a galaxy, there may be groups of stars, called star clusters, and galactic clusters may be part of what are even larger structures known as superclusters. Typical galaxies contain anywhere from 1 billion to 10 trillion stars; some of these stars may belong to star clusters. We live in one such galaxy, the Milky Way, which contains about 100 billion stars. The Milky Way is part of a small cluster, known as the Local Group, which in turn is part of a supercluster called the Virgo Cluster.

There are probably 100–200 billion — or even up to one trillion — galaxies in the universe, most residing in clusters and all far away from our own. The nearest large galaxy to us is the Andromeda galaxy, which is about two and a half million light-years away from us. This is almost 100 billion times the distance between the Earth and the Sun (one light-year is the distance light travels in one year, a huge distance indeed, since light travels at 186,000 miles per second and takes eight minutes and 20 seconds to reach Earth from the Sun). The Milky Way is the one galaxy we mostly see every clear night when we look up at the northern sky. On a moonless night, we can also glimpse a faint patch from the Andromeda galaxy. Astronomers believe that Andromeda is racing toward the Milky Way galaxy at 70 miles per second and will likely collide with it in about 5 billion years, eventually forming a larger galaxy.

Although the stars and galaxies appear almost still and steady when we look up at the night sky, they are in constant motion through the vast darkness of space. Being far away, their motion is hardly visible to the naked eye. Yet, constant motion is the profound truth of our universe. Planet Earth moves around the Sun once every year, and the Sun travels around the center of our galaxy once every 225 million years. As the Sun moves around, it wobbles back and forth across the galactic plane roughly every 30 million years, carrying with it the entire solar system. Every spiral galaxy like ours is spinning like a Frisbee as it runs away from every other galaxy, as time goes on.

As the constant motion in the sky continues, we may ask: Is it purely random, or does it follow certain rules? If it does, what are those rules? Almost a 100 years ago, the fundamental law that governs the movements of heavenly objects such as galaxies was written down by the great American astronomer Edwin P. Hubble. This is known as Hubble's law, which states that the speed of a star viewed from another star or planet depends on how far apart they are from each other. The farther apart they are, the faster the speed of one relative to the other.

With this knowledge comes a puzzle. Why doesn't this constant motion of galaxies tear a cluster apart, and why doesn't the constant motion of stars tear a galaxy apart? Why has the universe remained almost the same for more than 13 billion years?

We know that the Earth is constantly moving around the Sun, yet the force that keeps the Earth in a steady orbit around the Sun is the force of gravity exerted on it by the Sun. So, what is it that keeps the stars bound within the galaxies and galaxies bound within the galaxy clusters?

This fundamental question — what is holding the universe together — has puzzled thinkers for centuries. A simplified version of the above question is: "On what does the universe rest?" Thales of Miletus in Greece, in the seventh century BC, pondered this question and concluded that the universe rested on water. In the fourth century BC, the great Greek philosopher and polymath Aristotle suggested that it might be resting on ether. Hindu mythology held the belief that the universe rests on the back of four elephants standing on the back of a turtle. Other speculations have continued to arise over the centuries.

A more scientific answer emerged only less than a century ago with the discovery by Fritz Zwicky that there must be unseen matter in the galaxies — matter hidden from our view yet responsible for holding galaxies together. Zwicky was studying a galaxy cluster known as the Coma Cluster, located in the constellation Coma Berenices, a faint constellation nearly 300 million light-years away from the Earth. Though faint, it is still visible in a clear, dark night in the eastern sky.

Zwicky found that the galaxies in the Coma Cluster were moving so fast that the gravitational force of visible matter in the cluster failed to explain their speeds. He suggested that there must be hidden matter in the cluster (that he called *Dunkle Materie* or dark matter), which was many times more abundant compared to visible matter. This is discussed further in Chapter 2.

Astronomical observations since then have established that the dark matter that Zwicky found in the Coma Cluster is all-pervading. It is inside galaxies, outside galaxies, and everywhere in the universe. Its "pull" on ordinary matter has not only given birth to the universe we see today but has also provided the forces among the stars and galaxies, keeping them in such a delicate balance for billions of years. This state of balance is likely to continue for many billions of years to come, keeping the various parts of the universe interconnected in an unbreakable bond.

This fundamental discovery raised many questions: "What is dark matter? How can we find it? With what kind of force does it pull normal matter? Does it also pull or push itself with a force? Does it reside in our universe or elsewhere and pull familiar matter with its gravity? Are its manifestations clear enough to make it testable?"

We will explore these questions in the following chapters of this book, explaining why we think there could possibly be a second universe — called here "the replicate universe" — which is the home of dark matter, and what we may infer about the nature of this universe, if it is there, in the absence of any direct observations of its contents. How then can we learn more about this new universe if it exists? And what impact can it have on the future of the visible universe?

This book tells the story of a particle and a universe. The focus is on the dark matter particle and the question of where it resides: in our universe or in a new one that exists side by side with ours. The book is organized as follows.

Part I, after a brief historical interlude, describes how the existence of dark matter was established through astronomical observations. This is followed by the various theories of dark matter and a summary of experiments that attempt to find it. The failure of these experiments to detect the dark matter particle provides the launching pad for our hypothesis of a replicate universe (or mirror sector of the universe). It is argued that uncovering the laws of nature operating in this possible new sector of the universe requires an understanding of the standard model, which embodies the laws of physics governing the subatomic particles and forces in our universe. This is discussed in Chapter 3.

Part II explains how the laws relevant to the replicate universe can be deduced by applying the mirror symmetry. The main focus of this part is the detailed nature of how dark matter arises from the replicate universe, along with its astronomical and particle physics implications, which can be used to test this idea.

Part III focuses on the broader implications of this scenario. It is argued that if such a new sector exists, it will most likely resemble our own, with stars and galaxies much like those in our universe. The possible differences are highlighted, followed by speculation on their potential paleontological implications and an examination of what they could mean for the future of life on Earth. This part concludes with a speculative discussion of how, in the distant future, communication with this new "neighbor" might even become possible.

Part I

Chapter 1

Why Even Think of Many Universes? A Historical Perspective

> "It is not because things are difficult that we do not dare, it is because we do not dare that they are difficult."
>
> — **Seneca**

It is human nature to wonder about what lies beyond the boundaries of our familiar environment. This sense of inquisitiveness has compelled humans to travel to the Moon and send space missions to Mars, Jupiter, and beyond. It has also been at the root of age-old speculations about the existence of other universes and what that would imply for mankind.

The idea of the multiverse can be traced back to ancient Greece, where in the 4th century BC, a group of philosophers known as the atomists argued that all matter in the universe was composed of tiny, indivisible particles called atoms. These philosophers, among them Epicurus, reasoned that if it was possible to create one universe from such particles, then there must be other universes that had been created in the same way.

Aristotle and Ptolemy, on the other hand, proposed a geocentric model of the universe with Earth at the center and the stars and planets revolving around it. They argued that if there were an infinite number of identical universes, then no single universe could be regarded as the center of the cosmos. This would make the multiverse concept untenable. Geocentrism, and the resulting one-world view, persisted until the 16th century, when Copernicus showed that the Earth revolves around the Sun [1, 2].

The ancient Greek civilization was not the only one to grapple with questions about the nature of the universe. Centuries earlier, the Mayans had developed their own cosmological ideas. They believed that the universe consisted of three realms: the celestial upper world, comprising cosmic objects; the terrestrial middle world (Earth), where mortals lived, and a watery underworld, inhabited by the Gods. At the turn of the first millennium, the prolific and influential Persian philosopher and Sunni mystic, Muhammad Al Ghazali, entertained the idea of many worlds in his writings, arguing that God had created our world out of all possible worlds in which life and civilization could exist. The phrase "all possible worlds" presupposes the existence of many worlds [3].

By the end of the first millennium, the idea of the multiverse continued to circulate, even as the Church promoted a geocentric view of the universe. One of the earliest European scholars to consider the idea of many universes was the theologian and thinker Robert Grosseteste [4–6]. Born around 1170 AD in Stradbroke, England, to poor parents, he rose to prominence, becoming the Chancellor (probably the first) of Oxford University and eventually the Bishop of Lincoln. Grosseteste wanted to understand the nature of light and color. He was also a cosmologist. His research led him to suggest that the universe started out as a point of light and expanded to its present form. He closely analyzed Aristotle's geocentric theory in his book on light, *De Luce*, and discussed the possibility of one universe arising from many surrounding ones. This implicitly suggested the existence of many universes.

Meanwhile, the Catholic Church continued to promote the Aristotelian belief in a geocentric single-universe view. The Papal edict of 1277 banned any teachings to the contrary, declaring them heretical. Into this environment came the prominent 16th-century Italian thinker and intellectual Giordano Bruno, who openly advocated a many-worlds view in his writings. He published the book, *De linfinito universo e mondi* (*On the Infinite Universe and Worlds*), which presented arguments supporting this idea. By that time, Copernicus had presented his heliocentric theory, according to which Earth revolves around the Sun rather than the other way around. Bruno was inspired by Copernicus's work and postulated that the stars visible in the sky had planets around them and that some might even harbor life. This led to his many-worlds view, which was a direct challenge to the prevailing view held by the Church.

After lecturing abroad for many years about his theory, Bruno was invited to Venice by an influential patrician, Giovanni Moncenigo. Venice was, then, a very liberal city and far less dogmatic than many other places. Due to this, Bruno felt comfortable returning to Venice. His critical opinion of the Church, as well as his advocacy of many worlds, however, remained with him. This did not endear him to the Catholic Church. Eventually, his benefactor Moncenigo turned against him and denounced him to the Venetian Inquisition. Bruno then paid the ultimate price for refusing to abandon his views and not supporting the geocentric view. The Catholic Church had him burned alive in February 1600 in Campo de' Fiori in Rome. It was a tragic end to the life of an illustrious philosopher and cosmologist, whose thinking was far ahead of his time.

The Catholic Church continued to hold on to the geocentric view. One other person they persecuted was Galileo, who was using German astronomer Johannes Kepler's astronomical observations of Earth not being at the center but revolving around the Sun. Galileo was a contemporary of Bruno. He was sent to trial by the Inquisition and subsequently placed under house arrest for the remainder of his life for this belief. Even though by then, the scientific community and much of European society had accepted that the Earth was revolving around the Sun (and not the other way around), it was only in 1822, almost 200 years after Galileo's time, that the Catholic Church abandoned the geocentric view. In 1992, Pope John Paul II officially conceded that the geocentric view was wrong, and in 2000, he apologized for the way the Catholic Church treated Galileo. The pope also apologized for Bruno's execution but not for rejecting Bruno's ideas.

Older religions in the Orient, such as Hinduism, Buddhism, and Jainism, also had the concept of multiple universes built into their religious beliefs. For instance, Hinduism, which is more than 4,000 years old, advocated three worlds: heaven, where Gods and Goddesses live; Earth, where men and women reside, and hell, where sinners are exiled to. This belief persists even today, and ancient scriptures such as the *Ramayana*, *Mahabharata*, and the *Vedas* incorporated it into their teachings. Buddhist philosophy arose around 500 BC as a protest against various practices in Hinduism, such as the caste system, restrictions on women, and certain ritual practices. It was founded by Siddhartha Gautama, who was born into a wealthy noble family in Lumbini, India (now part of Nepal). He was disenchanted by old age, death, sickness, and other common human sufferings and rejected his life of riches, as well as his wife, Yashodhara, and son, Rahul, to seek enlightenment as a spiritual ascetic. Buddhists believe that there are 10 realms (or 10 worlds) through which a soul has to pass to attain enlightenment or Nirvana. This forms part of Buddhist cosmology. These states of existence could be interpreted as different worlds. Similar beliefs appeared in Jainism, which also started in ancient India around the same time as Buddhism. It also had a three-world belief similar to Hinduism and the Mayans: *Urdhva Loka* (upper world), *Madhya Loka* (middle world), and *Adho Loka* (lower world).

1.1. Quantum Physics, Extra Dimensions, and String Theory

In the 20th century, speculations about the existence of many universes began to appear in physics for entirely different reasons. Before the 20th century, physics dealt only with macroscopic systems, and their motion was described by the laws of Newtonian physics. Things, however, began to change around 1900 when physics underwent a revolution, following experiments conducted to study the motion of tiny atoms that make up macroscopic objects. A new set of rules called quantum mechanics had to be introduced to describe the behavior of atoms and molecules — rules that were drastically different from those of Newtonian physics. The old rules were, however, not completely discarded, but rather they emerged as limiting cases of quantum rules when large objects were considered. Since quantum mechanics was being used to describe completely different situations, new assumptions had to be introduced to get a full understanding of its rules. In this context, the concept of the multiverse was reintroduced, as discussed in the following section.

1.1.1. *Quantum Mechanics*

Broadly speaking, physics deals with understanding the time evolution of the states of a system. What is a "system," what is a "state," and what is its "evolution"? Any object we are studying, be it a piece of chalk, a car, a cloud in the sky, a volume of gas, or an apple falling from a tree, is called a "system." The "state" of the system (which could be one of the above or any other systems) describes what condition the system is in at a given instant in time. For instance, if a moving car is the system under consideration, its state can be described as follows: The car, weighing 3,000 pounds (its weight), is in Baltimore (i.e., its location), moving at a speed of 30 miles per hour towards Washington DC (its speed). This tells you everything you need to know about the car at that instant. You could, of course, ask other questions, such as how many passengers are in the car or how many of them are male or female. You could add this information to the above information on the state if needed. Collectively, this information describes the state of the system.

The "time evolution" of the system (a moving car in Baltimore) can be described as follows: After an hour, the car arrives in Washington, DC, i.e., the state of the car evolves from being in Baltimore at 10 AM to arriving in Washington, DC, at 11 AM. Time evolution requires using a clock to keep time. The system must be followed minute by minute or second by second, depending on how precise the information needs to be. The resulting information can be recorded in a graph of the location of the car on the road versus time from the beginning to the end of the trip. This graph can be a straight or curved line, depending on whether the car is moving at a constant speed or at different speeds.

In order to have precise knowledge of the location of the car at different moments as it moves from Baltimore to Washington, someone must observe it from start to finish. The act of observation is intimately connected to the time evolution of a system. Observation can be made by someone in another car, driving side by side, or by a drone flying overhead, taking a movie of the car, or by a driver sending text messages of the car's location every few seconds. It is through these observations that we gain knowledge of the detailed state of the system (car) and its time evolution. Since the systems are large, the act of observation, which may involve sending a beam of light at the car and detecting when it is reflected, doesn't affect the position or speed of the car, and its time evolution remains unaffected. However, this is not the case for atomic systems.

To determine the location of an atomic particle, we can follow the same procedure. In this case, however, two things happen. First, when a beam of light hits the particle, the power of the beam will jostle it into another position, and you can't tell where exactly the particle is by detecting the reflected light. This causes uncertainty about its position. Second, the beam transfers energy to the particle, thereby also changing its speed, making it difficult to ascertain its exact speed. Of course, if the particle moves faster, these effects are less pronounced. Thus, in contrast with macroscopic systems, the act of observation introduces a new element of uncertainty into the discussion of the motion of atomic particles.

By the 1920s, it was becoming clear that atomic systems required a different set of rules to study their time evolution due to the uncertainty introduced by the act of observation. These rules, developed in the early part of the 20th century by Niels Bohr, Werner Heisenberg, Erwin Schrödinger, Arnold Sommerfeld, and Max Born, are known as quantum mechanics. One of these rules, proposed by Heisenberg, is called the Heisenberg uncertainty principle. Heisenberg showed that uncertainty in the position of a particle caused by the act of observation is related to the uncertainty in its speed, as already seen. This is because the faster an atom moves, the less its position is disturbed by the incident beam of light. This uncertainty principle became a fundamental element of the laws of quantum mechanics. Heisenberg received the Nobel Prize for this work in 1932. The citation read *for the creation of quantum mechanics, the application of which has, inter alia, led to the discovery of the allotropic forms of hydrogen.*

The German physicist Max Born proposed that in an atomic system, we could only know the probability of a particle's position and speed at a given instant. This probability is encoded in a mathematical description, called the wave function, introduced by Erwin Schrödinger in 1926. The wave function for a system contains all the information about the state of the system, such as its speed and position at a given time. Each state has a different probability. Schrödinger gave an equation, known as the Schrödinger equation, which describes how the wave function evolves with time. Given the wave function of a system, determining which particle state a particle is in requires an act of measurement. When the system is measured, its state becomes definite even though the wave function contains information about all the possible states. This would appear to be a contradiction. How can this be?

Heisenberg and Bohr pondered this question. In 1927, they postulated that a separate assumption must be added to quantum mechanics, namely, that the act of measurement selects one state from the many states contained in the original wave function. They called this phenomenon the "collapse of the wave function." They believed that the collapse of the wave function was due to the interaction of the atomic system with the measuring device. This interpretation came to be known as the Copenhagen interpretation of quantum mechanics.

1.1.2. *From the Collapse of Wave Function to Many Universes*

The collapse of the wave function was an alarming new concept in quantum mechanics and rightly raised many profound questions. For example, after the collapse, what happened to the other states of the system that were not observed? Did they just disappear once one state was observed, or were they still there after the measurement? What would happen if one attempted to measure another state of the same system subsequently? Understandably, wave function collapse became a highly contentious topic and drew the attention, in subsequent decades, of luminaries of quantum physics, such as Erwin Schrödinger, John von Neumann, Eugene Wigner, Freeman Dyson, and Albert Einstein. Einstein, who was famously skeptical of quantum mechanics, is said to have remarked, "God does not play dice."

What is puzzling about the wave function collapse is beautifully illustrated by the so-called "Schrödinger's cat paradox," developed by Erwin

Schrödinger in 1935 in a discussion with Albert Einstein. Schrödinger suggested a thought experiment in which we envision the wave function of a cat in a box with a vial of poison and a radioactive triggering device inside. The vial may or may not be triggered to open by the radioactive device. If a radioactive event occurs, the vial is shattered, spilling the poison that kills the cat; on the other hand, if the radioactive event does not happen, the cat is alive. Thus, the wave function describing the cat in the box is a superposition of two states, a dead cat and a live cat, depending on the probability of the radioactive event occurring. But once the box is opened, the wave function of the cat collapses, and the observer can determine whether the cat is alive or dead, but not both at the same time.

In 1957, a young physicist named Hugh Everett III proposed a resolution to this conceptual puzzle. Everett was completing his PhD thesis at Princeton University under the renowned physicist John Wheeler. Everett, the son of a World War II lieutenant Colonel from Washington DC, had changed his undergraduate major from chemical engineering to mathematics before pursuing physics as a graduate student. He proposed in his PhD thesis that when a measurement is carried out on a quantum system with many states, all the states would appear in the outcome. This is different from the Copenhagen interpretation, which holds that only one state appears. Everett suggested that each state would appear in a different universe and that we would observe only the state present in our universe. Each universe would have its own observer who would see a different state. Everett had conceptualized a multiverse within the context of quantum mechanics [7]. In Schrödinger's thought experiment, where the wave function is a superposition of two states — a dead cat and a living one — an observer in one universe sees the cat alive, while an observer in another universe sees the cat dead. The act of observation splits up the universe, turning it into a multiverse. This revolutionary concept raised new questions. Did the resulting universes reflect reality? Could one travel from our universe to any of the other universes, or communicate with them? It appeared that there were no answers to such questions and that there never would be. There was no way to test such ideas experimentally.

The multiverse idea of Everett was not received well in Copenhagen. This was partly due to the radical nature of the suggestion and partly due to Bohr's strong commitment to the idea of the wave function collapse. Everett's thesis advisor, John Wheeler, who had worked with Niels Bohr earlier, visited Bohr at the Institute for Theoretical Physics in Copenhagen in May 1956 [8] after mailing a draft of Everett's thesis to him and his associate Aage Petersen. He was unsuccessful in convincing Bohr of the validity of Everett's theory. Then, in the spring of 1959 [8], at Wheeler's insistence, Everett requested an interview with Bohr and was granted one. But when they met, Bohr was not interested in listening to Everett's ideas [9]. The visit, according to Everett, was a "complete disaster" [10]. Leon Rosenfeld, another one of Bohr's associates, called Everett "indescribably stupid," adding that he "could not understand the simplest things in quantum mechanics" [10].

Frustrated by the rejection of his idea and worried that he would lose his military draft deferment [8], Everett switched from physics to weapons research and joined the Weapons System Evaluation Group (WSEG) in the US Department of Defense. He did not return to physics after that [11]. Meanwhile, the lack of a satisfactory explanation for the collapse of the wave function lingered as a major conundrum in quantum mechanics. Only toward the end of Everett's life did the multiverse begin to gain traction, seemingly providing a philosophical way out of this conundrum [11].

1.2. Extra Space Dimensions

Quantum mechanics was not the only context in which the idea of a multiverse arose in physics. It appeared independently in discussions of the possibility that, unbeknownst to us, there might exist hidden space dimensions in nature. The idea was first entertained in 1919 by the Polish physicist and mathematician Theodor Kaluza. Until that time, it was thought that space had three dimensions, which we can visualize as follows: When an airplane flies, it goes forward, sideways, and upward, corresponding to the three space directions. Einstein unified the three space dimensions with time in his theories of special and general relativity. Kaluza wanted to extend Einstein's theory of general relativity to include an extra space dimension.

According to Newton's theory of gravity, two massive bodies such as the Sun and Earth exert an attractive force on each other — gravity — that keeps Earth revolving around the Sun in an elliptical orbit. Einstein's general theory of relativity instead envisioned gravity as a curvature of space–time around a massive body. According to this interpretation, the space around the Sun becomes curved, and Earth moves along the shortest path in this space — an arc — which creates an elliptical orbit. The extent of the curvature depends on how massive the object is. This interpretation of gravity in terms of geometry was a revolutionary idea that attracted the attention of many physicists, including Kaluza.

The only other force known at that time was that between electric charges: the electric force. Kaluza wanted to determine whether that force could also be related to the curvature of space in a manner similar to gravity. He noticed that if he added an extra space dimension to Einstein's theory of general relativity, he could derive the electric force between charged particles from the curvature of this extra dimension. By using the mathematical trick of adding a fourth space dimension, he had developed a unified theory of two fundamental forces of nature — gravity and electricity. He was, of course, well aware that there was no evidence for a fourth space dimension. There still is none. Despite this, the idea was so exciting that he immediately wrote to Einstein, who, after some initial hesitation, sent it for publication in the *Proceedings of the Prussian Academy of Sciences* in 1921. With Einstein's sponsorship, Kaluza's work became widely recognized.

A few years later, Oskar Klein, a Swedish physicist, followed up on Kaluza's work. Klein completed his doctorate in physical chemistry at University College, Stockholm, Sweden. Before defending his thesis, he spent a year at Copenhagen working with Bohr and became interested in physics. After receiving his PhD, he moved to the University of Michigan in 1923, where he took a position as Professor of Physics. In 1925, at the encouragement of Hendrik Kramers, an associate of Bohr, he moved back to Copenhagen, which at the time was the center of theoretical physics. While at Michigan, he had become interested in extra dimensions [12].

While Kaluza had assumed that an extra space dimension was only a mathematical entity, Klein assumed it to be real and gave it a quantum interpretation. The pressing question became: If extra dimensions were real, why hadn't anyone seen evidence for them? Klein's answer was simply that the extra dimension was curled up and too tiny to be visible. A crude illustration of how extra dimensions curl up is shown in Fig. 1.1. Suppose you take a plain sheet of paper, which has two dimensions, and roll it up until it becomes infinitesimally narrow like a thin wire. The rolled-up paper looks like it has only one dimension, with the second dimension almost completely invisible (Fig. 1.1). The curling up of the extra dimension is called compactification.

Klein's quantum interpretation of the curled-up extra dimensions had far-reaching implications. According to quantum mechanics, if a particle is confined to a box of a certain size, its energy is inversely proportional to the box's size, i.e., the smaller the size, the higher the energy. The same logic, when applied to extra dimensions, implies that a particle moving in an infinitesimally small dimension would have a high-energy state. Klein argued that such high-energy states could be detected experimentally, indicating the existence of extra dimensions. Today, these high-energy states are called Kaluza–Klein states. Klein summarized his arguments in a short paper published in *Nature* in 1926. Despite the novelty of the Kaluza–Klein theory, these two papers received little attention until the development of string theory, in which they provided a new motivation for extra dimensions. More recently, two pioneering papers revived this field [13] by suggesting how extra dimension models

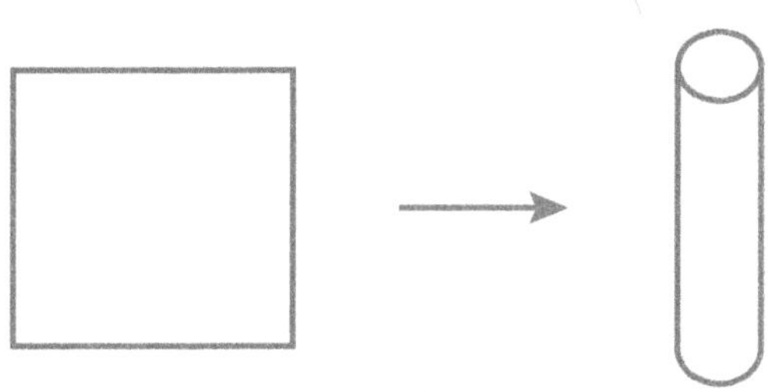

Fig. 1.1. Curling up of one of the two dimensions of a plain sheet of paper.

can be applied to resolve problems with the standard model of particle physics (for an explanation of the standard model, see Chapter 3).

To see how extra dimensions imply the possibility of multiple universes, think of two walls separated by a distance in familiar three-dimensional space. The walls are two-dimensional, and two ants crawling on the two walls can travel only in two directions. For the ants on each wall, the world is two-dimensional. The two worlds inhabited by the ants are connected by a bridge in the third dimension, and this setup could be considered as two separate universes.

In 4+1 extra-dimensional models, one can imagine analogous situations where there is a line in the fourth space dimension connecting the two three-dimensional volumes at its ends, much like the walls in our previous example. One could imagine each of the volumes as a separate universe. This will then lead to a picture of two universes in five space–time dimensions. A schematic illustration is shown in Fig. 1.2. According to this picture, there are two island universes at the end of an extra space dimension. Being in one universe, we would not see the other one, but only feel its effect due to the gravitational force exerted by matter in that universe.

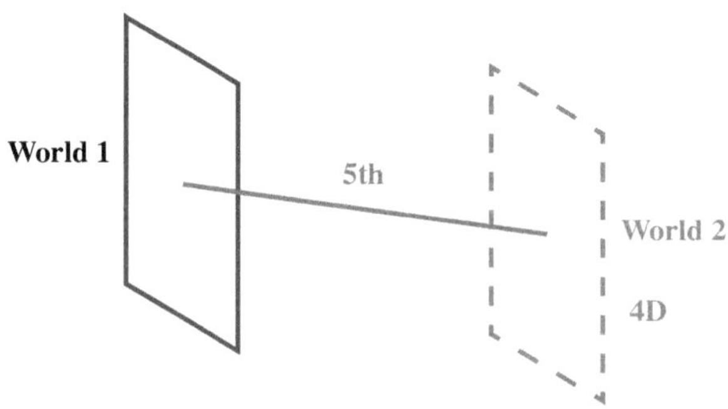

Fig. 1.2. Two universes (World 1 and World 2) in an extra-dimensional model. The two rectangular regions represent the two four-dimensional (three space + one time) regions in a five-dimensional (four space + one time) world. Each four-dimensional world could be regarded as a separate universe, just like the two walls of the ants in the familiar space–time.

1.3. String Theory and Many Universes

All matter, be it a brick, a piece of chalk, or a house, when broken down to its ultimate pieces, leaves a heap of dust as its last remnant. Each dust particle looks like a dot on a piece of paper, having a tiny mass but no dimension, and is called a point particle in science. Physicists from the time of Galileo and Newton to Bohr and Einstein assumed that all matter that we see consists of such point particles at the most fundamental level. They were the ultimate constituents with forces acting on them and holding them together to make matter. During the last century, it was established that forces are also caused by the exchange of similar but different point particles between the matter constituents. Thus, the fundamental role of point-like particles in understanding natural phenomena became widely accepted in science. In the 1970s, however, a completely new approach was suggested according to which forces and matter were postulated to arise instead from the motion of extended string-like objects.

In the late 1960s, there was no complete theoretical framework for describing the behavior of the various forces and matter. The string approach arose from attempts to understand some of the forces better and was initiated by several physicists, including Gabriele Veneziano in Europe, Yoichiro Nambu and Leonard Susskind in the US, followed by significant contributions in the 1980s by John Schwarz and Michael Green at Caltech and David Gross and Edward Witten at Princeton.

A string left to itself constantly wiggles and is known to create waves of different wavelengths, much like a vibrating string on a violin. According to string theory, these waves represent both matter and the force particles. Quantum mechanics had already revealed that particles can be described by waves and vice versa; therefore, waves on strings describing particles like electrons and protons were not an unfamiliar idea (Fig. 1.3).

String theory obeys strict mathematical rules, which are different from those governing the motion of point particles. These rules not only imply that extra space dimensions must exist in the universe but also, more interestingly, predict the number of space–time dimensions to be

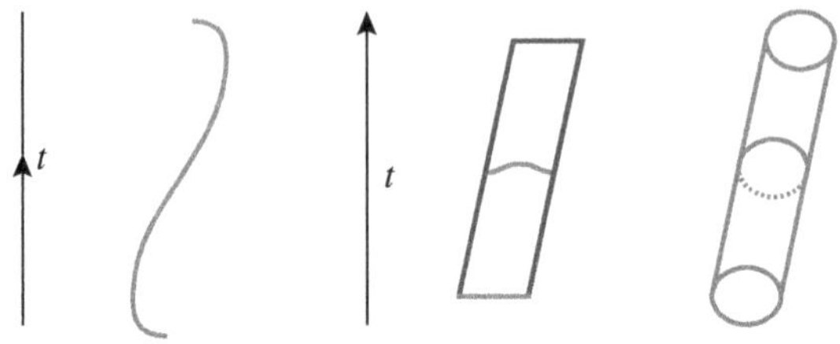

Fig. 1.3. Propagation of a point particle, an open string, and a closed string. *t* denotes time and arrow denotes direction of time.

either 10 (one time + nine space) or 26 (one time + 25 space), thus providing a rationale for Kaluza–Klein-like theories. From the work of Oskar Klein, we knew that having extra space dimensions is not a fundamental obstacle to building models of the universe. The extra space dimensions in string theory (six in the first case and 22 in the second) must be curled up to tiny sizes so that the familiar four-dimensional space–time emerges out of them. The curling can lead to new forces of nature, just as electric forces arise from the curling of the fifth dimension, as demonstrated by Kaluza. Curling more than one extra dimension in string theory is more complex, and in this process, all known forces in nature can emerge.

Recall the curling up of one dimension of a plain sheet of paper with two dimensions, as shown in Fig. 1.2. There is only one way to do this so that the space looks like an ultra-thin tube with one visible dimension. When there are two extra space dimensions, they can be curled up into two independent ways: as two pencil-like curlings or as a doughnut shape (see Fig. 1.4). This picture would then imply two universes: one associated with the pencil-like curling and another with the doughnut-type curling. When there are more dimensions to be curled up, there can be myriad ways of doing it, leading in each case to a different possible multiverse picture in the familiar four-dimensional space–time. Each universe will have its own laws governing matter and forces. All the universes that emerge this way will inhabit the same space but will be different from each other. Our universe is just one of them and has the right conditions to obey the laws of physics that we are familiar with. There may be other universes where this is also true.

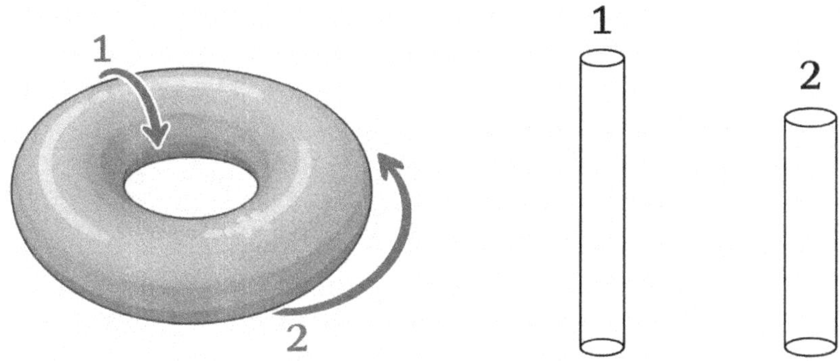

Fig. 1.4. Two extra dimensions and two universes: one in the left panel and the second on the right. Numbers 1 and 2 denote the two extra space dimensions that have been compactified.

The first question one can ask about this picture is: "Can one experimentally test the multiverse predicted by string theory?" We do not know the answer to this question. It is quite possible that once higher-energy colliders are built, they may be able to probe deeper into the hidden world of extra dimensions and detect the new Kaluza–Klein particles predicted by the theory. So far, however, no such evidence has been uncovered.

The multiverse picture of string theory has clearly another unsatisfactory aspect to it due to the possibility of a huge number of universes it predicts. It leads to the second question: "How could the single universe we are familiar with arise out of the many universes predicted?" Attempts have been ongoing to add extra assumptions to string theory to achieve this goal. One approach has been to invoke a principle known as the anthropic principle, which states that our universe must have properties that allow the emergence of intelligent life so that observers can ask questions about the universe [14–16]. Demanding the anthropic principle would then pick out the real one. This is, however, not very satisfactory because it raises the question: Why should we build an elaborate apparatus such as string theory if, in the end, we still must invoke an extra *ad hoc* principle to explain the known universe?

In conclusion, speculations in history and mythology, as well as scientific reasonings in various contexts, have made the idea of many universes not only a matter of intellectual curiosity but also a conceivable and potentially realistic possibility, making it an attractive subject for scientific research.

Are there any other indications for such a many-worlds picture? We will see that the conclusive evidence for dark matter in astronomy, which will be explored in the ensuing chapters, makes a compelling case for two universes, providing an additional incentive to probe the multiverse idea. To better appreciate why the existence of dark matter might imply this, we need to develop a full picture of what is already known about familiar matter and forces. This will provide a stepping stone to our continued exploration of dark matter and its connection to many universes.

Chapter 2

What Holds the Universe Together?

> "Our knowledge can only be finite, while our ignorance must necessarily be infinite."
>
> — **Karl Popper**

As the 19th century drew to a close, many important developments had already taken place in science. Darwin's theory of evolution by natural selection, outlined in his 1859 book *On the Origin of Species,* had been widely accepted in biology. Significant developments in chemistry included the development of atomic theory by John Dalton in 1803, the birth of organic chemistry through the work of Friedrich Wöhler in 1828, and the formulation of the periodic table of elements by Dmitri Mendeleev in 1869. In physics, laws governing the motion of charged particles, namely, electricity and magnetism, had been established during the 18th and 19th centuries. Planetary motion was understood using Newton's laws of gravity proposed as early as 1687. There was no evidence for any new kind of fundamental force beyond gravity, electricity, and magnetism. Collectively, these laws were called the Newtonian laws of physics. Most physicists were beginning to conclude that all laws of nature governing matter had been discovered.

However, this sense of completeness was soon overturned by two major discoveries made in the closing years of the 19th century: the observation of beta ray emission from nuclei by Henri Becquerel in 1896 and the discovery of X-rays by Wilhelm Conrad Röntgen in 1895. These were followed in 1900 by Max Planck's revolutionary idea of discrete light quanta, suggesting that light is emitted in discrete packets rather than continuously. Einstein's groundbreaking paper on relativity followed 5 years later. Physics was poised to enter its most transformative era!

Other new discoveries were also being made around that time regarding matter. Toward the end of the 19th century and the beginning of the 20th, scientists started looking at what is inside the atom and how things work in the subatomic world. The first clue came from the work of J.J. Thomson. In 1897, he discovered that electrons were inside atoms by observing rays in a cathode ray tube, which consisted of gas at very low pressure confined between two metal plates at high voltage. Since the rays were coming from the plate with negative voltage, the cathode, he concluded using the well-understood laws of electricity that they consisted of tiny, negatively charged particles repelled by the cathode. He called them *electrons* and pointed out that they must reside inside the atoms that make up the cathode. Thus, the first constituent of an atom was discovered.

Attention then shifted to determining the other constituents of the atom. There was a two-pronged strategy to unravel this mystery.

First, since atoms are electrically neutral, it was correctly inferred that there must be positively charged particles inside the atom to balance the negative charge of the electrons. The positively charged particles were called *protons*. The protons and electrons in the atom would then be held together by the attractive electric force and move around each other, similar to planets orbiting the Sun.

Second, since the laws of electricity implied that the accelerated motion of charged particles would lead to the emission of light, one way to understand the inner workings of atoms was to study the light emitted from them. In the atom, as electrons spiral around the proton, light would be emitted, and it was initially believed that the light spectrum would consist of continuous colors. However, when the atomic spectra of different atoms were studied by Swedish physicist Ånders Angström in the 1850s, he found that different atoms emitted light of different colors, and these spectra were discrete rather than continuous. Further studies by John Balmer in 1885, Theodore Lyman in 1906, and others confirmed this. The discreteness of spectra was a surprise and was believed to be related to atomic structure, but nobody knew how.

To probe this further, scientists sought to determine how protons and electrons were distributed inside atoms. Were the positive charges distributed like a plum pudding with electrons embedded within it, as J.J. Thomson suggested, or was it different?

To test Thomson's theory, Rutherford assigned his student Ernest Marsden to check whether positively charged alpha rays shot at a gold foil would be deflected backward or continue forward as the plum pudding model predicted. Marsden found that many alpha particles were scattered backward [17]. This simple observation amounted to an unexpected and momentous discovery that surprised Rutherford. He interpreted it to conclude that the atoms had a dense central core that repelled the incoming positively charged alpha particles backward. He called this core the nucleus of the atom and suggested that it was surrounded by electrons in orbits around it like planets around the Sun. In a further experiment, when Rutherford bombarded nitrogen nuclei with energetic, positively charged alpha particles, he observed what appeared to be a hydrogen nucleus emerging from the nitrogen. He called it the proton. Protons were found to be much heavier than electrons. Now, we know that the size of the nuclear core is less than a trillionth of a centimeter and that of the atoms is about 1,000 times larger. Although the existence of the proton was conclusively established by Rutherford, almost two decades after the discovery of the electron, its existence had already been suspected earlier by Eugene Goldstein in 1886.

Rutherford and later Chadwick noted that helium was heavier than the weight of two protons, which it should have to balance the charges of two electrons. This implied that there had to be something almost as heavy as a proton, but without electric charge inside the helium nucleus. Chadwick initially believed that since the extra particles had no electric charge, they might be a proton and an electron joined so tightly that they behave like a single electrically neutral particle, with their charges canceling out. In principle, this would also explain why Becquerel found beta rays, or electrons, emitted from the nucleus. However, this turned out not to be the case. Instead, physicists found that the nuclei of helium and other atoms heavier than hydrogen contained another elementary particle, which had no electric charge; they called it a neutron. This hypothesis was confirmed when the neutron was discovered as a single particle ejected from nuclei. The correct picture of the atom was established: The nucleus consists of protons and neutrons at the core, surrounded by very light electrons circling around. The plum pudding model of Thompson was disproved.

The neutron took some time to be discovered since it does not have any electric charge and hence is not attracted or repelled by charged plates. British physicist James Chadwick discovered it in 1932 by scattering energetic alpha particles on a beryllium nucleus, which emitted an electrically neutral particle. When this particle was directed at a hydrocarbon target, it produced detectable signals. This particle was the neutron, and Chadwick was awarded the Nobel Prize for this discovery in 1935.

As the atomic spectra were studied further by Ångström, Balmer, Lyman, and others, it became clear that the well-known 19th-century laws of electricity could not explain the discrete nature of the spectra. A new set of rules had to be developed for this purpose. The first major breakthrough in understanding this came from the works of Niels Bohr, who proposed a completely new framework for light emission by atoms. This gave birth to the field of *quantum mechanics*.

The laws of quantum mechanics governing atoms were novel and path-breaking, and were slowly developed during the first three decades of the 20th century. They provided an understanding of all atomic phenomena known at the time. In particular, they were remarkably successful in explaining the colors of light emitted by atoms observed in experiments.

It was known since 1910 from Einstein's work that the light emitted by atoms consisted of particles called photons. The name *photon* was given by chemist Gilbert Norton Lewis in 1926 in a letter to nature. Thus, the four fundamental particles known at that time were protons, neutrons, electrons, and photons.

2.1. Zwicky Discovers *Dunkle Materie*

All these developments from 1900 to 1930 successfully explained almost all experimental observations of the time and appeared to provide a complete set of working rules for the subatomic world. This made physicists feel that practically everything there was to know about matter had been learned.

This could have been, therefore, another moment similar to that of 1900, when it was felt by most physicists that all matter and forces in the universe had been discovered and there was nothing more to be learned. There were certainly physicists who felt that way at the time. But into this atmosphere of overconfidence came the observation by Fritz Zwicky in 1933, which shook the scientific community. He argued that protons, neutrons, and electrons (and photons) were not the only fundamental constituents of the universe. Zwicky claimed that he found evidence for a new kind of matter in galaxies.

Zwicky was a Swiss-American astronomer, born in Bulgaria to a Swiss father, Fridolin Zwicky, and a Czech mother, Franceska Vrcek, in 1898. After obtaining a PhD from the Federal Institute of Technology in Zurich under Peter Debye, he moved to the US in 1925 to work with the Nobel Prize-winning physicist Robert Millikan at the California Institute of Technology. He remained there for the next 50 years, until his death in 1974. He initially worked on solid-state physics and gaseous ionization, but later turned his attention to astronomy. He persuaded the California Institute of Technology to build him a telescope equipped with the recently invented Schmidt camera, so that he could photograph large numbers of galaxies. Using these new telescopes, he studied different galactic clusters, including the Coma Cluster, which is located in the northern sky and is more clearly visible to the naked eye in the spring months. He focused in particular on measuring the speeds of galaxies within this cluster.

To determine the speeds of galaxies, he analyzed the color of light they emitted and measured its wavelength. He then compared this wavelength with the known emission spectra of atoms. If a source of light is moving away from an observer, the wavelength of the emitted light increases, causing its color to shift from its original value toward the red or even infrared region, depending on how fast it is receding. The smaller the recession speed, the smaller the shift in the wavelength. Conversely, if the star is moving toward the observer, the light wavelength becomes shorter, and the color shifts toward the blue or violet end of the spectrum. This phenomenon is known as the Doppler effect. It applies not only to light but also to sound waves and has played a fundamental role in both astronomy and physics. It also has important applications in medicine, for example, in Doppler imaging techniques. This effect is named after its discoverer, the 19th-century Austrian mathematician and physicist Christian Andreas Doppler.

Doppler was born to a family of stone masons in Salzburg, Austria. He graduated with a degree in Mathematics from the University of Vienna in 1829 and struggled to secure a stable academic position of his liking. After various temporary appointments, he finally landed a regular position in Prague in 1841, just as he was about to finalize his plans to immigrate to America. He decided to remain in Prague, and in 1842, he presented his work on the frequency change of waves due to the motion of the source or the observer in a treatise entitled *Über das farbige Licht der Doppelsterne und einiger anderer Gestirne des Himmels* (*On the Colored Light of the Double Stars and Certain Other Stars of the Heavens*).

He proposed his idea to explain the shift in the frequency observed from rotating binary stars. His theory could also be applied to other kinds of waves. For example, it explained why sound appeared less shrill (lower frequency or lower pitch) when the source is moving away and more shrill (higher pitch) when it is moving toward the listener; the faster the source moves away, the lower its pitch, and vice versa. Doppler realized that as the source of sound or light moves, the wavelength of the wave is stretched or compressed, depending on the direction of motion: when moving away from a listener, the wave is stretched, and when moving toward the listener, it is compressed (see Fig. 2.1). In the case of sound, whose speed has a constant value in air and is given by the product of its frequency and wavelength, an increase in wavelength corresponds to a decrease in frequency (resulting in a lower pitch), and vice versa. A similar effect occurs for light as well.

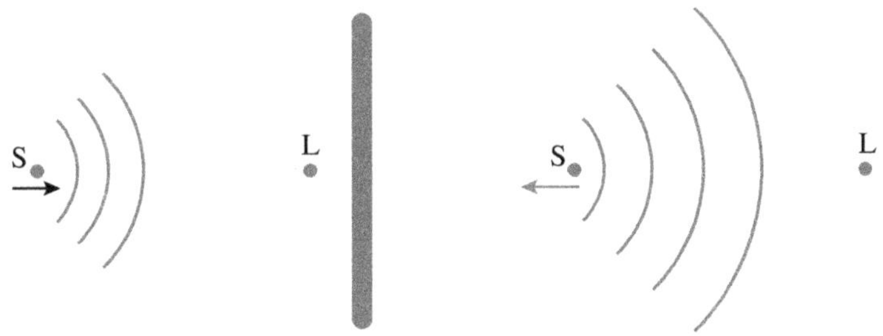

Fig. 2.1. Doppler effect of sound and light waves when the source or observer is moving. Arrows next to S denote the direction in which the source is moving. Using Hubble's law and the Doppler shift, astronomers established that all stars and galaxies in the universe are in constant motion, receding from one another at great speed.

When Doppler presented his theory of wavelength shift, he was ridiculed and vehemently criticized by a well-known mathematician, Joseph Petzval, who argued that all natural laws should follow from a differential equation rather than from simple algebraic manipulations used by Doppler. Petzval proposed a competing principle called "the conservation of oscillation time in undulatory phenomena," which was diametrically opposed to the Doppler effect. According to Petzval, the frequencies of oscillation-generated waves will never change regardless of the motion of the source. Petzval was so forceful in his opposition to the Doppler effect that mild-mannered Doppler was overwhelmed, and his discovery could not gain a foothold in physics during his lifetime. In 1852, a meeting of the Austrian Academy was convened to hear the arguments both for and against Doppler's theory. The meeting was well attended, and Petzval argued vigorously in favor of his own ideas. Unfortunately, Doppler was unable to attend the meeting as he was suffering from tuberculosis and traveling to Italy for treatment. Petzval succeeded in convincing the audience of his point of view, and the final vote was overwhelmingly in his favor; Doppler's theory was effectively dismissed. Four months later, Doppler died in Venice, Italy, in 1852, where he had gone in search of a cure for tuberculosis. It was only after his death that his groundbreaking work received full recognition, aided by subsequent contributions from scientists, such as the British civil engineer John J. Scott Russell, the French scientist Hippolyte Fizeau, and the Austrian physicist Ernest Mach [18].

The Doppler shift was also used by Edwin Hubble to establish his groundbreaking rule of celestial motion, known as Hubble's law. This law relates the speed of galaxies to their distance from the observer. He used the Doppler shift of light from distant galaxies to calculate their receding speed and determined the distances by using a class of stars called Cepheid variables, whose intrinsic brightness was related to their pulsation periods. The brightness of an object goes down with distance, much like a candle that becomes dimmer as we move farther away from it. This relationship helped Hubble to establish the distance of a galaxy in which the Cepheid is located. Cepheid variables were discovered before Hubble by Henrietta Leavitt. Hubble analyzed the ratio of recession speed to distance for 46 galaxies and found it to be constant. This proportionality constant is now known as the Hubble constant and is an extremely important parameter in cosmology.

The same Doppler effect was also used by Zwicky, who found in his study of the Coma Cluster that the galaxies at the periphery of the cluster were moving too fast and yet not flying away. Since it is gravity that keeps galaxies bound to the cluster, the question arose: "Is the gravitational force of visible galaxies in the cluster strong enough to keep such fast-moving galaxies bound within the cluster?" To understand this using a familiar analogy, consider a rotating turntable with wet clay balls stuck at its edge. As the turntable rotates slowly, the clay balls remain attached because the frictional force between the disk and the clay balls is sufficiently strong. However, as the disk rotates faster and faster, there comes a point when the clay balls fly off because the frictional force that held them at lower speeds is no longer sufficient. In the case of the Coma Cluster that Zwicky was observing, the analog of the frictional force is the gravitational force from the mass of visible matter (i.e., stars and galaxies) in the cluster. The strength of this gravitational attraction can be estimated from the number of stars and galaxies present: More stars and galaxies imply greater mass and hence a stronger gravitational force. There are about 1,000 galaxies in the Coma Cluster, and each galaxy has billions of stars. Using this information, Zwicky estimated the total mass in the cluster, which allowed him to calculate how strong the gravitational attraction on an outer galaxy would be. Zwicky found that the mass required to hold the fast-moving galaxies in the

Coma Cluster was almost 100 times greater than the total mass of visible galaxies estimated above. He therefore concluded that something must be missing from his estimate of the mass in the cluster. Since he could not see anything else that could provide the additional mass needed, he proposed that there must be some unseen matter in the Coma Cluster contributing to the gravitational force of the visible mass of the stars and galaxies. He called this unseen matter *dunkle Materie* or dark matter.

This was clearly a fundamental discovery, but at that time, atomic physics was the prime focus of physicists, and something unknown in outer space that Zwicky claimed to have observed was not taken very seriously. The Second World War came a few years later and placed the survival of human civilization in doubt, slowing scientific progress. It was only after 1945, when the war ended, that life began to return to normal and scientific research regained momentum. Physicists refocused on understanding the atom better, and Zwicky's groundbreaking discovery remained merely a footnote.

Zwicky's observation, however, would go on to shake the fields of physics and astronomy nearly half a century later. To appreciate the significance of his discovery, let us recall the nature of visible matter, namely, protons, neutrons, and electrons. These particles have the important property that they can be seen with our eyes or with scientific instruments, such as a telescope pointing at the sky or a microscope looking at minute samples. The basic principle behind these commonly practiced ways is the use of light. Light is either emitted from an object or reflected from it, allowing it to be observed. This is how we see the stars and galaxies in the sky and tiny microscopic organisms in the laboratory. Of course, we cannot directly see electrons, protons, or neutrons with the naked eye, since they are too small to reflect sufficient light; however, we can detect their presence indirectly using devices that use electricity or magnetism. For example, when these tiny objects pass through plates coated with light-sensitive chemicals used in early photographic techniques, they leave clearly visible tracks that can be analyzed. They work the same way as a camera used to take our photographs. Modern digital photography using computers, such as those used in smartphones, of course does not need such chemicals. The tracks on light-sensitive plates provide clear evidence of the existence and motion of these tiny particles, such as protons and electrons.

Eventually, photographic plates were replaced by more advanced detection devices. For example, particle detectors called cloud chambers were invented by Scottish physicist Charles Wilson in 1911, which made it possible to observe the tracks left by the tiny subatomic particles as they passed through the chamber, knocking electrons off the medium. Their electric charge would break up other atoms and knock off their electrons, producing visible tracks. The track would bend and curl, depending on the electric charge of the initial particle. In this way, the existence of particles could be inferred, even though they were not directly visible. It can also tell us what kind of particles they are. Depending on the direction in which the tracks bend, we can determine the sign of their electric charge and estimate their mass. Heavy particles bend much less than the lighter ones, while positive and negative charges bend in opposite directions. By analyzing where the tracks originate and how frequently they are produced, scientists can also gain insight into the nature of forces involved in their creation.

Although neutrons do not carry electric charge, they can be detected using devices that contain materials such as lithium, helium, or some other elements that absorb neutrons and produce protons afterward. These end products can then be observed in the particle detectors described above. In some cases, photons (or light) are also produced, which can, of course, be detected directly. Thus, scientists developed a variety of techniques to observe subatomic particles that constitute the visible world.

A profound question that emerged from this knowledge was: Is the matter we can observe — either directly or through detector tracks — all the matter that exists in the universe, or is there some other form of matter that remains invisible to these devices? Zwicky's discovery cast serious doubt on the prevailing belief that protons, neutrons, and electrons constituted all matter. The ultimate question then became: If Zwicky was correct, where does dark matter fit into the scheme of things in the subatomic world? This question remained unanswered then, and to a large extent, remains so even today. However, numerous theoretical proposals have been put forward in recent decades, and experimental efforts to detect dark matter have increased dramatically. What followed resembled a form of scientific detective work — one that continues to this day and is likely to do so for many years to come.

As a historical footnote, Zwicky was not the first to suggest the existence of dark matter. Indications of hidden forms of matter in the universe had appeared in the literature long before his 1933 paper. As early as 1877, Father Angelo Secchi, a Catholic priest and astronomer from Italy and Director of the Roman College Observatory, referred to "dark masses" in his study of properties of nebulae [19]. He also made important contributions to astronomy, including the study of comets and a class of hot, bright stars called Be stars. In the latter half of the 19th century, physicists such as Lord Kelvin also suggested the possible existence of dark objects (or gas) in the Milky Way galaxy [19], based on studies of stellar motion. However, it appears that Lord Kelvin may not have taken his own observations very seriously or pursued them further. Similar conclusions about invisible forms of matter were also drawn in the early years of the 20th century by Estonian astronomer Ernst Öpik [19] in 1915, Dutch astronomer Jacobus Kapteyn [19] in 1922, and Jan Oort [19] in 1932. All these studies, however, were not as compelling as Zwicky's work. He was the first person to present convincing scientific evidence that there must exist an unknown form of matter beyond what we see around us. For this reason, he is often regarded as the "father of dark matter."

2.2. Rubin and Ford

The next phase in the exploration of dark matter came with the monumental work of two American scientists: Vera Rubin and Kent Ford. Vera Rubin was born in Washington DC as Vera Cooper in 1928. She developed an interest in observing the night sky full of stars from her childhood. She received her degree in astronomy from Vassar College in 1948, but was denied admission to Princeton's graduate program because she was a woman. Instead, she joined Cornell University for her graduate education, where she studied under distinguished physicists such as Richard Feynman, Hans Bethe, and Philip Morrison. She later obtained her PhD degree from Georgetown University, working under George Gamow, who was then at the nearby George Washington University. She subsequently secured a permanent position at Carnegie Institution in Washington, where she collaborated with Kent Ford to study the motion of stars in galaxies. Kent Ford, who was born in Virginia, received his PhD from the University of Virginia and later joined the Carnegie Institution, where he developed advanced instruments for astronomical observations. Rubin and Ford used these instruments to measure the speed of stars in galaxies.

Working at Kitt Peak Observatory, Rubin and Ford spent years observing distant stars through a 36-inch telescope to study stellar motion in different galaxies. They began with the Andromeda Galaxy, measuring the rotation speed of stars as a function of their distance from the center of the galaxy. To achieve this goal, they analyzed the light emitted by stars in the galaxy and measured changes in its frequency (or color) using the Doppler effect as the stars moved around the center. This allowed them to determine the speed of the stars.

The stars farther away from the center of the galaxy are expected to move more slowly, since gravity is weaker the farther you move from matter. It is just like the motion of planets in the solar system, where planets farther from the Sun move more slowly. However, Rubin and Ford found that the speeds of stars remained constant, no matter how far they are from the center of the galaxy. This surprising result led them to conclude that there must be some unseen matter (dark matter) distributed throughout the galaxies, exerting extra force on stars farther away and helping them maintain their speed. They estimated the amount of dark matter required for this purpose and found, much like Zwicky, that the total dark matter mass in the galaxies was about 10 times the visible mass. The revolutionary scientific significance of Rubin and Ford's work was profound: There was now much stronger direct evidence for dark matter in the universe (see Fig. 2.2), reinforcing and extending Zwicky's earlier findings.

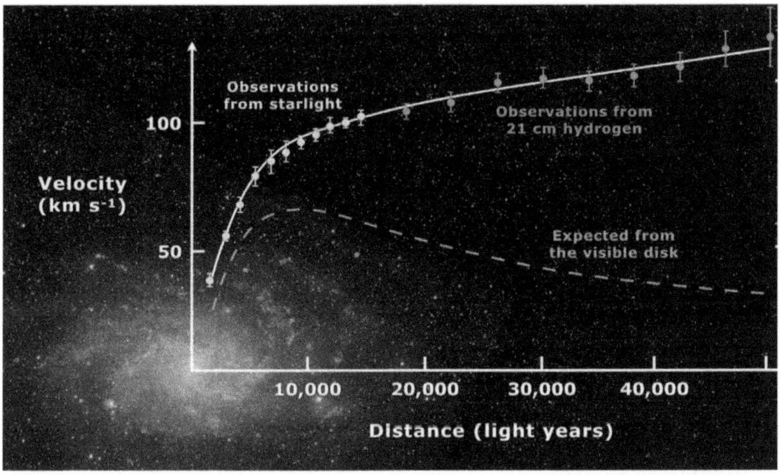

Fig. 2.2. The flat rotation curves observed by Rubin and Ford provide strong evidence for the presence of dark matter in galaxies (created by Mario de Leo, Wikipedia).

Vera Rubin was eventually awarded the National Medal of Science by Bill Clinton. She did not receive the Nobel Prize for her pioneering work.

Other astronomical evidence has since appeared that corroborates these findings. One of the most striking examples is the observation of Bullet Cluster by the Chandra X-ray Observatory (CXO), named after the famed Nobel Prize-winning Indian American astrophysicist Subrahmanyan Chandrasekhar. The Chandra space telescope was launched aboard the Space Shuttle *Columbia* in 1999 and has made many important observations. The observation relevant to dark matter is that of the collision of two galactic clusters, known as the Bullet Cluster. This system was formed from the collision of two large clusters of galaxies and is located 3.8 billion light-years away. From observations of this collision, astronomers were able to infer the presence of dark matter in the cluster. The reasoning is as follows: visible matter interacts not only through gravity but also through electromagnetic (Coulomb) forces of the charged particles within it. This is why ordinary matter experiences resistance, for example, we cannot push our hand through a concrete wall. In contrast, dark matter does not interact with a

Coulomb force and therefore passes through visible matter much more easily. In Fig. 2.2, the regions corresponding to visible (familiar) matter appear as the brighter, more concentrated areas, while the dark matter is represented by the more diffuse and extended regions. Since dark matter does not exert an electric force on itself or on ordinary matter, it is not significantly slowed as it passes through visible matter in a galaxy. In contrast, visible matter experiences drag due to the electric force. As a result, in the cluster collision, dark matter is ahead of normal matter, which lags due to the dragging effect.

Another way that dark matter makes its presence felt is via a phenomenon called gravitational lensing. Gravitational lensing is similar to optical lensing, which we are all familiar with from the eye doctor's office and from optics laboratories. Consider an optical lens, such as the one used in eyeglasses. When light rays from an object pass through the lens, the lens bends (refracts) the rays and focuses them onto the

Fig. 2.3. Image of the Bullet cluster, showing the spatial separation between visible matter and dark matter. Brighter, denser regions represent visible matter, while more diffuse regions represent dark matter. The latter experiences negligible interaction during the collision and therefore moves ahead of visible matter, which is slowed by electromagnetic interactions, leading to a clear separation between the two.

retina of the eye. The bending makes the source appear brighter and clearer. A similar effect occurs when light passes near a massive object, such as a star, a planet, or a distribution of dark matter. The mass exerts a gravitational pull on the light, bending its path and producing a lensing effect similar to that of an optical lens. This can be seen as fuzzy patches in the Bullet Cluster (see Fig. 2.3) when looked at carefully. Those fuzzy, broadened patches of light are the gravitational lensing effect of dark matter in the two colliding galaxies. This phenomenon was also first predicted based on Einstein's theory of general relativity and was later emphasized in the context of galaxies and clusters by Zwicky.

2.3. Dark Matter and Galaxy Formation

Dark matter also helped resolve another major mystery of the Big Bang model of the universe. As the Big Bang model was developed, cosmologists sought to understand how structures such as galaxies and stars formed in the universe. According to the Big Bang theory, the universe started off as an extremely hot fluid in a tiny volume and has been expanding and cooling ever since. Within 10,000th of a second after the Big Bang, protons and neutrons formed. About a minute later, they combined to form atomic nuclei, primarily hydrogen and helium, along with small amounts of lithium and deuterium. About 100,000 years after that, electrons combined with these nuclei to form neutral atoms. This picture of the universe presented a puzzle. According to the theory, the expanding hot cosmic fluid of atoms was uniform and homogeneous. However, our universe is anything but homogeneous; rather, it is filled with non-uniformities — empty spaces interspersed with dense pockets of matter in the form of galaxies, planets, stars, and other celestial objects. So, how did such non-uniformities arise from an initial uniform fluid? Dark matter resolved this puzzle in a very simple way. According to this theory, as the universe expanded and cooled, dark matter was moving slowly, and chunks of it formed dense pockets at various locations in the universe. This happened about 30–100 million years ago. These pockets are known as *halos*. The gravitational pull of these halos attracted visible matter, causing it to accumulate and form increasingly dense clumps. Over time, this process led to the formation of galaxies, beginning roughly about a billion years after the Big Bang.

Detailed observations and simulations have confirmed this picture. Thus, without dark matter, the formation of galaxies would be extremely difficult to explain.

2.4. Assessing Hidden Matter

All indications are that dark matter — an invisible form of matter in the universe — is far more widespread than was recognized prior to the 1930s. As noted earlier, the key property of this dark and unseen matter is that it does not reflect or give out any light, which is why it remains invisible, unlike stars and planets that either emit or reflect light. Dark matter, however, exerts gravitational forces on other matter and thus plays a crucial role in holding the visible matter in galaxies and galactic clusters together.

So, what kind of matter is dark matter [19]? Could it simply consist of massive objects made of protons and neutrons that do not shine? Or is it a completely new kind of particle, analogous to protons, neutrons, and electrons, except that they do not shine? Alternatively, could it be composed of extended, diffuse, and invisible structures that are spread across the universe? Let us examine these possibilities.

Belonging to the first category are objects known as brown dwarfs. These are composed of protons, neutrons, and electrons, but they do not shine. One may ask how objects made of protons, electrons, and neutrons can fail to shine. The Sun is made of protons, neutrons, and electrons, but it shines. On the contrary, the Earth is also made of the same constituents, but it does not shine. This is due to the fact that when we have compact massive objects like the Sun or the Earth, their total weight determines whether they shine or not. When the weight is high, like the Sun, the pressure of the gravitational force at the center is so high that the temperature there goes way up, making the proton and electron undergo nuclear fusion. In this process, they release energy, which makes stars like the Sun shine. Since Earth does not have as much mass, its central temperature is lower; therefore, nuclear fusion that takes place inside the Sun does not occur inside the Earth, preventing it from shining. There are objects in mass between the Sun and the Earth called brown dwarfs, which are more massive than Earth but not heavy enough like the Sun to shine. The existence of such objects was first proposed by Indian physicist Shiv S. Kumar in 1962. About 2,800 brown dwarfs have been identified by now, including one close to the Sun. They could, in principle, qualify as dark matter since they do not emit any light. It has, however, been established that there are not enough brown dwarfs to account for the required amount of dark matter, even within our own galaxy.

Thus, it appears that dark matter is a completely new form of matter, beyond anything we are familiar with. It is not yet known whether it experiences new kinds of forces of non-gravitational nature either with itself or with known matter. Identification of this new form of matter will be a fundamental breakthrough in science. For this reason, there has been intense interest in the subject, and many experimental efforts to detect dark matter particles are currently underway.

We do not yet know whether dark matter constitutes only one kind of particle or multiple kinds of new particles. Nor do we even know whether it is composed of fundamental particles or more complex, extended structures.

2.4.1. *How Much Dark Matter Is There in the Universe?*

Scientists have attempted to answer this question in several ways. One approach is to determine how much extra mass is required to explain the constant rotation speeds of stars in galaxies as a function of their distance from the center (Fig. 2.2). Similarly, one can ask how much extraneous mass is needed to account for the subtle features observed in the cosmic microwave background. What emerges from such research is a striking result: Almost 84% of all matter in the universe is dark matter, whereas only about 16% is visible matter in the form of stars, planets, and interstellar gas.

In addition to visible and dark matter, there exists another form of energy that fills the universe, known as dark energy. Evidence for this component has emerged only in the past few decades. It arises from observations indicating that the expansion of the universe is currently accelerating. This acceleration cannot be explained by normal matter or dark matter, both of which can lead only to a slowing down of the expansion of the universe. This is because the rate of expansion of the universe depends on matter density, which decreases as the universe expands. The accelerated expansion must therefore come from a new form of energy whose density does not decrease, i.e., an all-pervading form of energy that we cannot see. This is referred to as dark energy. An overview of the present-day energy composition of the universe is shown in Fig. 2.4.

Figure 2.4 indicates the amount of dark matter in the universe. In addition to remaining unseen, this places important constraints on the properties of dark matter particles. For instance, dark matter cannot be arbitrarily heavy, since otherwise, unless its abundance is very small, its content right now will exceed the 26% of the energy content shown in Fig. 2.4.

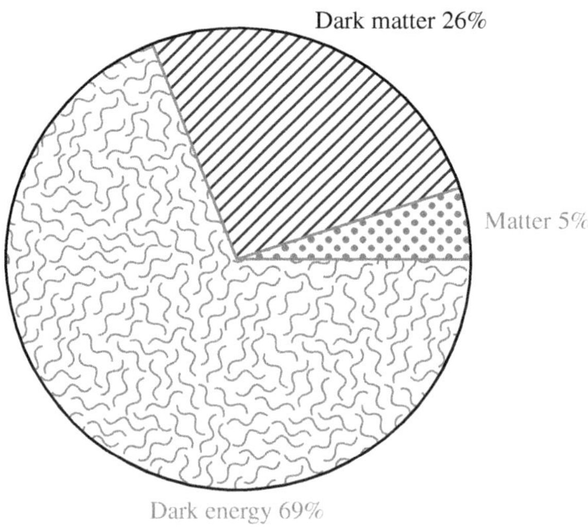

Fig. 2.4. Energy content of the universe: 69% dark energy; 26% dark matter; 5% visible matter. Dark energy is a uniformly distributed component responsible for the accelerated expansion of the universe, whereas dark and visible matter contribute to gravitational attraction and slowing down of the Hubble expansion.

It is reasonable to expect that dark matter must also be undergoing changes as the universe evolves, like light waves, and whatever changes it experiences in the early universe must be consistent with the observed present-day abundance, ensuring that its contribution remains close to the 26% indicated in Fig. 2.4.

We do not know how much dark matter, if any, was there at the very beginning of the universe. It could have been abundant, or it may have been entirely absent and produced later as the universe evolved. If dark matter were very abundant in the early universe, it would exceed the observed amount today unless the excess part "dies away" as the universe expands. To "die away," it must annihilate or decay to known particles by just the right amount so that it remains precisely 84% of the universe's current mass density. This requirement places constraints on the properties of dark matter, particularly on how it interacts with known matter. If dark matter did not exist in the beginning but since the known particles have to be there, then it must have been produced through interactions of known particles. Again, to match the presently known amount, the known particles must interact with dark matter in a precise way. In either case, these new restrictions on how dark matter and known matter act on each other can be exploited in experimental searches for dark matter particles in the laboratory.

It is therefore important to understand visible matter as clearly as possible. Fortunately, there exists a well-established theoretical framework describing visible matter and their interactions with each other. This is known as the standard model, which has been extensively confirmed by experiments. Our understanding of visible matter, based on the standard model, can then become a springboard for different ideas on dark matter particles, thereby providing a platform for analyzing the physics behind it. We therefore discuss the essential features of the standard model theory of visible matter in the following chapter.

Chapter 3

Particle and Forces on Space–Time: The Standard Model

> "A model is only the face of a hidden reality that you are able to imagine."
>
> **— Anonymous**

> "...any physical theory is essentially only a model for the world of experience, we must renounce all hope of finding anything like the correct theory... simply because the totality of experience is never accessible to us."
>
> **— Hugh Everett**

The autumns in Rochester, New York, are cool harbingers of the harsh winters to follow. It was no different in 1966, when I arrived there from India after completing my master's degree to work with two of the giants in particle physics: Robert E. Marshak and Susumu Okubo. Getting used to the new environment took time. After some preliminary coursework, I was ready to start serious research with them in the summer of 1967. Winter in Rochester is especially uncomfortable for someone who comes from a tropical country. Spring, however, brought with it good weather and hope. With renewed energy and enthusiasm, I started participating in group seminars, trying to find an exciting topic to work on. The hard part came pretty soon afterward. Even though I had some familiarity with the field, finding a suitable topic for research was not easy. Particle physics in those years was going through a lean period. A lot was known about forces and matter at the subatomic level, but there was no organized approach to put them together and hence no set program for a graduate student to begin working with.

The 1960s were a time when the study of the universe was in its infancy, and important questions about it, such as dark matter, were not even on the "radar screen" of physicists. Cosmologists were focused only on a preliminary understanding of the evolution of the universe. Visible matter occupied the attention of particle physicists. They were searching for ways to better understand the structure of matter and the forces that we could see and feel every day. A practical reason for this was that scientists had gathered enough information about visible matter and its interactions. They were trying to put together that information to develop a fuller understanding of nature.

This journey to understand visible matter goes back many centuries to the ancient Greek philosopher Leucippus of Miletus and his disciples, such as Democritus (5th century BC). They suggested that all matter is built out of indivisible blocks called atoms. This idea was developed over the years till the dawn of the 20th century, when it became clear that behind the vastness that we see around us were two simple concepts: matter and forces in total embrace — forces holding matter together and matter exerting forces on other matter to stay in place. Out of this symbiotic dance between matter and forces were born the nuclei that made the stars shine, the atoms that bind to form molecules, and the molecules that bind to give the primordial dust that made planets like our own around stars. The question then was: What is the detailed nature of these subatomic forces that make this possible? Advancing our understanding to a deeper level became the central task to which physicists devoted themselves in the mid-20th century and beyond. As very ambitious students, we wanted to be part of that history and possibly contribute to it.

3.1. Matter: Quarks and Leptons

As noted in Chapter 2, it was already realized during the first decades of the 20th century that all matter is made of three elementary particles: protons, neutrons, and electrons. The proton has a certain amount of electric charge. We call that amount one unit or simply +1. The neutron is electrically neutral (i.e., it carries no electric charge), and in the same units, the electron has an electric charge of −1, exactly opposite to the proton charge. The law of electric forces discovered by Charles Coulomb between 1785 and 1787 revealed that the positive charge of a proton attracts the negatively charged electron to form neutral atoms. This is much like the force of gravity that keeps the Earth orbiting the Sun.

Protons and neutrons with nearly the same mass are the heaviest of the three basic constituents. They reside in the core of an atom, called the nucleus, and the electrons, about 2,000 times lighter than the proton, are in orbits at a tiny distance around the core. The fact that protons and neutrons stay together at the core of an atom was established by groundbreaking experiments conducted by Ernest Rutherford, as discussed in Chapter 2.

This raised two immediate questions: since the protons have a positive charge and since, according to the laws of electricity, like charges repel each other, how are the protons in an atomic nucleus able to stay together, instead of pushing each other away and breaking up the nucleus?

A second mystery arose when the protons and neutrons were found to weigh nearly the same. They were two different particles with different electric charges, and *a priori,* it was not clear why their weights should be almost the same.

The first puzzle was solved by postulating that there is a new kind of attractive force acting on the protons at the atomic core, which is strong enough to overcome the electric repulsion force between them. This force must also attract both protons and neutrons so that they can all be confined to the core of the atom without breaking apart. More about this force is discussed in the following.

Trying to understand what fundamental principle of nature caused the protons and neutrons to have the same mass was much harder, but its understanding eventually unlocked the secrets that led to the foundation of modern subatomic physics. The first step was to determine whether only protons and neutrons have the same mass, or whether there are other groups of particles with a similar pattern of nearly equal masses. This way, one could determine how general this property is and whether it reflects something fundamental.

To study this, scientists invented high-energy collision machines, called accelerators, which generated beams of high-energy protons and electrons, so that they could probe inside the atom. Those higher-energy particles were then made to hit the proton head-on to see what is inside them. In these collisions, they not only broke apart the atoms and revealed what is inside but also produced new particles in the process.

Another source of high-energy particles is cosmic rays from outer space that are constantly bombarding the Earth. In cosmic rays, energetic particles collide with atoms and molecules in the atmosphere. In this process, new particles also appear, providing further richness to our knowledge about new particles and nuclear forces.

Particles called π-mesons, K-mesons, and muons, all with different properties from protons, neutrons, and electrons, were discovered in cosmic rays. Of these, the muon was later established to be a cousin of the electron since, like electrons, it does not experience nuclear forces. The electrons and muons are called leptons. On the other hand, π-mesons and K-mesons were more like protons and neutrons, and participated in exerting nuclear force. They were followed by many other groups of particles with near-equal masses like protons and neutrons, and all of them exerted nuclear force, for example, the three π-mesons with charges (+1, 0, −1) have nearly the same masses as the K-mesons with charges (+1, 0). This implied that the general principle that made protons and neutrons have nearly equal mass applies to more particles. These new particles, as well as other new ones discovered since then, are listed in a biennially published particle data book from Lawrence Berkeley Laboratory, called *Particle Data Tables.*

The first proposal to understand the closeness of the masses of protons and neutrons came from Werner Heisenberg, who suggested that there is a symmetry called isospin that flips a proton to a neutron, and vice versa. The nuclear forces that bind protons and neutrons respect this symmetry. This guarantees nearly equal masses for them as well as for groups of strongly interacting particles, such as pions and kaons. That was the beginning of the era of symmetries in particle physics.

The exploding particle list and how to understand the properties of these particles became the challenge facing scientists in the 1960s. Then, in the mid-1960s, came a breakthrough: Murray Gell-Mann of Caltech and George Zweig of Geneva suggested a systematic way to understand the exploding particle list. They proposed that these particles were composed of more fundamental constituents, much like protons, neutrons, and electrons are the constituents of atoms. Thus, protons and neutrons were understood not to be truly elementary particles.

The particles inside protons and neutrons were called quarks. Two kinds of quarks were needed to explain protons and neutrons: they were named u (for up) and d (for down) quarks. They had unusual values for their electric charge: u had charge 2/3 and d had charge −1/3. No free quark has been discovered with such fractional values for electric charge, raising another puzzle, whose resolution remains an open question. Regardless of this, however, it became possible to understand both protons and neutrons using combinations of three u and d quarks. Protons were made of two u quarks and one d quark, while neutrons were made of two d quarks and a u quark. This made the proton have an electric charge of +1 and the neutron have zero charge, as observed. If the u and d quarks had nearly the same mass, then protons and neutrons, and separately the different π-mesons, which were made of an up quark and its anti-particle together with the down quark and its anti-particle in different combinations, would have nearly the same mass. Anti-particles are introduced later on in this chapter. The question of why protons and neutrons had nearly the same mass thus reduced to the question of why u and d quarks had nearly the same mass.

A third quark, the strange quark "*s*," was introduced by Gell-Mann and Zweig to understand the K-meson and its cousins. The strange quark was very similar to the down quark in that it carried the same amount of electric charge, but it differed in that it had a peculiar characteristic called strangeness. The up and down quarks had no strangeness. The K-meson was made of a strange quark "*s*," together with up and down quarks, and their anti-particles. They behaved differently compared to π-mesons because of the presence of a strange quark in it.

Since the K-meson was heavier than the π-meson, the strange quark was assumed to have a larger mass than the u and d quarks. Since then, the discovery of new but different particles in more energetic colliders has necessitated the introduction of three more quarks known as charm (c), bottom (b), and top (t) quarks, all heavier than the strange quark. With these six quarks, all known new particles could be understood.

Similarly, in the lepton sector, two more leptons — one more charged lepton (the tau lepton) and one more neutrino (the tau neutrino) — were discovered, making a total of six leptons. Thus, six quarks were accompanied by six leptons. This pointed to a symmetry between quarks and leptons.

Another fact about particles that has been known since the early 1930s is that, corresponding to every particle, there is an anti-particle. The existence of anti-particles was first suggested by the work of Paul Adrien Maurice Dirac, who discovered the concept while trying to extend the rules of quantum mechanics to the domain of fast-moving relativistic particles. Dirac's theory suggested that every particle was accompanied by an anti-particle, which was a completely new kind of object, weighing the same as the corresponding particle and carrying opposite electric charge. If a particle and its anti-particle meet at a point, they cause a fiery explosion, leaving no trace of the original particle or anti-particle. According to Dirac's theory, the electron would have an anti-particle with the same weight and opposite electric charge (+1), and if it comes near an electron, it will destroy the electron as well as itself. The anti-particle of the electron was called the positron and was discovered in cosmic rays by Carl Anderson in 1932. He received the Nobel Prize for this discovery in 1936. Even though Dirac's original theory only predicted an anti-particle for the electron, his logic applied to all particles, implying that all particles have corresponding anti-particles.

The quarks, leptons, and their anti-particles became the new fundamental particles of nature, replacing protons, neutrons, and electrons. Scientists could understand all the particles discovered in colliders and cosmic rays by putting together quarks and anti-quarks in different ways. For example, the positively charged pion has as its underlying constituents a *u*-quark, which has charge +2/3, and an anti-*d* quark, which has charge +1/3.

In addition to the quarks and leptons just mentioned, another particle that pervades the universe and plays an essential role in the building of the universe as we see it today is the neutrino. We are swimming in a sea of neutrinos, and every second, billions of them come from the Sun and pass through us. The neutrino has no electric charge, nor does it exert any strong force; so, we don't feel anything as it passes through us. The neutrino is also a lepton since it does not bind to protons and neutrons inside the atomic nucleus. It is only emitted in a beta decay process and participates in a very weak force. It turns out that each charged lepton is accompanied by its own neutrino.

This summarizes our current knowledge of the constituents of matter. We will call quarks and leptons, including their anti-particles, collectively matter particles.

Most physicists believe that there probably are other new kinds of visible matter particles, likely heavier, which will appear on the stage of space–time as we delve deeper into the subatomic world, armed with higher-energy colliders. This belief is so strong, and the arguments behind it are so compelling that the search for new forces and particles is the subject of intense research in particle physics now and will continue far into the future. Albert Einstein's famous theory of relativity predicts a relation between mass and energy, which tells us that we need machines with higher energies than the masses of the heavy particles to produce those heavy particles. If heavier matter particles exist, they have not been seen yet because the energy generated by the current machines is lower than their mass. There are plans to build particle colliders with much higher energy as new ways to accelerate particles are invented. The Large Hadron Collider (LHC) is the highest energy machine in the world today. It collides protons with protons at a total of 14 tera-electron volt energy, which is 14,000 times the mass of the proton (one tera-electron volt is a 1,000 times the mass of the proton). Plans are underway to increase this to 100 tera-electron volts at an accelerator being contemplated in China. When that happens, our ability to uncover new matter and forces will have taken another giant step forward indeed.

3.2. Forces and Their Origin

Let us now turn our attention to the properties of various forces acting among particles.

The first question one can ask is: How do forces between particles arise? Until the early part of the 20th century, forces were added to the discussion of physics problems as an extra assumption. But things changed drastically in the mid-1930s. While discussing the origin of nuclear forces, Hideki Yukawa proposed that the exchange of a particle, later identified as a *pion*, between protons and neutrons is responsible for this force. This introduced a drastic change in our thinking about forces. It suggested that, just as particles represent matter, there are also other kinds of particles responsible for forces. How that happens can be understood as follows: Imagine two people playing ball; when one person throws the ball to the other, the receiving person feels a force from the impact of the ball. Thus, the ball is the transmitter of the force exerted by one person on the other. In the language of atoms, the two persons playing ball represent the matter particles exerting a force on each other, and the ball is the force particle. The mass of the exchange particle determines how far the influence of the force extends. The heavier the exchange particle, the shorter the range of the force, and vice versa. This also squares well with common intuition since a heavier ball, when thrown, will travel a shorter distance than a lighter one.

The pion was discovered soon after Yukawa's suggestion, confirming this idea of particles being responsible for forces. Thus, it was realized that a single basic concept, the particle, underlies both matter and the forces we observe in nature. This provided an appealing unity to our understanding of the universe. We will see in the following section how such a simple picture provides a unified understanding of all observed phenomena in nature.

3.2.1. *Force Particles vs. Matter Particles*

How do force particles differ from matter particles? There is an intrinsic property of particles called spin that distinguishes between force and matter particles. Spin was a new idea that had to be added to the already existing rules of quantum physics. It originated from efforts to understand the nature of light emitted from atoms in more detail. The concept of spin was introduced by Dutch physicists George Uhlenbeck and Samuel Goudsmit in 1925 and is an attribute of all particles. It is a subtle concept. One can imagine it to be like a ball spinning about its axis, although, as often remarked, inside particles, "there is neither a ball nor any spinning." It is an intrinsic property of particles, governed by the rules of quantum mechanics, which allows their values to be discrete. These allowed values of spin are either integral, like (0, 1, 2,...), or half-integral, like (1/2, 3/2,...).

Matter particles have half-integral spins, whereas force particles have integral spins. The half-integral spin particles are called fermions after the famous Italian physicist Enrico Fermi, and integral spin particles are called bosons, named after Indian physicist Satyendra Nath Bose.

Satyendra Nath Bose was born in Calcutta (now Kolkata), India, in 1894 at the height of British rule. His father, Surendra Nath Bose, was an accountant with East India Railways, and his mother, Amodini Devi, was a homemaker. Despite working under the British government, Surendra Nath was a nationalist and opposed to British rule. Satyendra Nath Bose graduated from Calcutta University at the age of 19. He then taught at the University of Dacca (now in Bangladesh) before moving to Kolkata. Like his father, he was also a participant in the Indian National Movement opposing the British Raj. His main interest was, however, physics, and he contributed a great deal to our understanding of physical phenomena. His most important contribution was to point out how a collection of photons, the light quanta, behaves. This was later generalized to other integer-spin particles as well. All such particles are called bosons after Bose's last name. He, however, did not get the Nobel Prize despite having done this groundbreaking work, which was followed up by Einstein and describes the properties of half the particles in the universe. He was nominated for the Nobel Prize several times [20].

Among the fundamental force particles, only those with spin zero and spin one have been discovered in the laboratory. Similarly, we have not seen any elementary matter particles beyond spin 1/2 yet. Just like matter particles, force particles also have their anti-particles.

The following sections describe the four different forces of nature and their properties.

3.3. Gravity: The Oldest Known Force

Gravity was the earliest discovered force of nature. The laws of gravity were formulated by Isaac Newton. Born in 1643 in Woolsthorpe, England, Newton was a student at Cambridge University. In 1665, when the university closed due to an outbreak of plague, he moved to his parents' home in Woolsthorpe. It was during this period that he discovered the laws of gravity. He is said to have been inspired by observing an apple fall from a tree, leading him to connect terrestrial motion (falling objects) with celestial motion (planets orbiting around the sun).

Newton's laws of gravitation dominated the physics world for more than two centuries. A key first ingredient underlying this framework is the so-called equivalence principle. This principle states that if one drops a heavy mass (say a rock) and a light mass (say a feather) from the same height, they will hit the ground at the same time. This is contrary to common sense, and Galileo Galilei is said to have tested this principle around 1590 by dropping a heavy and a light object from the top of the Leaning Tower of Pisa in Italy. This principle appeared in Newton's theory of gravity and helped relate the motion of diverse objects, like the planets orbiting the Sun, to the apple falling from the tree.

This led him to conclude that gravitational force acts the same way on all objects, regardless of their composition. According to Newton's law, the gravitational force between two objects depends only on how massive the objects are and how far they are from each other. The force becomes weaker as the two objects are pulled farther apart and becomes stronger as the masses of the objects increase.

The equivalence principle was used by Einstein in developing his theory of gravitation, which superseded Newton's theory and became one of the cornerstones of 20th-century physics. More precise tests of the equivalence principle are being carried out even now [21], improving upon earlier tests conducted first by Galileo and later by Hungarian physicist Loránd Eötvös in 1908, followed by many others.

Newton's laws of gravity provided the foundation for the field of planetary science prior to the development of Einstein's theory. Using this, British physicist Edmund Halley could predict the timing of solar and lunar eclipses, and also the distances from the Sun to various planets from the knowledge of their orbital periods.

We could also combine the laws of gravity with the laws of heat to explain why life exists on some planets and not on others. A planet in the solar system receives heat from the Sun, which gives energy to the molecules in the atmosphere, enabling them to escape from the planet's gravitational pull. How far a planet is from the Sun determines how much heat it receives and how fast the molecules in its atmosphere are escaping. On the other hand, the gravitational force of the planet is trying to pull the molecules and atoms in the atmosphere toward itself. The atmospheric gas molecules, therefore, end up in a tug-of-war between a planet's gravitational pull and the heat from solar energy. If solar energy wins, the atmospheric molecules fly away, depleting the atmosphere over time, and if the planetary gravity force wins, the atmosphere stays in place. So, to have the right environment for life and a full atmosphere, there must be a delicate balance between these two forces. This is sometimes called the "Goldilocks Zone" around a star. Obviously, this is only a necessary condition, and other conditions are needed to generate and sustain life (e.g., water). We are lucky enough to have the Earth in a Goldilocks zone so that there is life. For the Moon, on the other hand, this balance has not worked out due to its smaller size and weaker gravitational force, leading to solar energy winning over gravity in depleting the atmosphere. Consequently, no life is possible on the Moon.

Einstein's theory of gravity, which superseded Newton's theory, had as its main thesis that gravity need not be viewed as a force but rather as a curvature of space. The presence of matter curves space–time. The more the matter, the more the curvature. As legendary physicist John Archibald

Wheeler put it, "matter tells space–time how to curve, and curved space–time tells matter how to move." Slowly, Einstein's new theory of gravity came to be accepted over time, and its various implications were studied. No deviation from Einstein's theory has been found.

Einstein's theory not only reproduced Newton's laws in situations when the gravitational field was weak but also predicted the Big Bang expansion of the universe, which is now assumed to describe how the universe evolved from its beginning. It also predicted black holes, gravitational waves, and many other phenomena, which have all been discovered, placing Einstein's theory on a firm footing.

Einstein's laws of gravity received their first spectacular confirmation in 1919 during a solar eclipse, when it was found that light rays from stars passing by the Sun were bent by exactly the amount that Einstein's theory predicted, but Newton's law did not. This observation was carried out by Arthur Eddington and Frank Watson Dyson, then Director of Cambridge Observatory, in Príncipe, an island off the coast of West Africa. Einstein's theory stated that a massive object curves the space around it, and in that curved space, objects follow the shortest path. This path would bend light traveling near the object. The path followed by an object near a massive object is called a geodesic. No additional mention of the force of gravity is needed to describe particle motion. By this logic, one would expect the space around the Sun to be curved due to the massive gravitational force of the Sun. A star that is behind the Sun, and thus blocked from the straight line vision of an observer, can now become observable because its light follows a geodesic in the curved space around the Sun (Fig. 3.1).

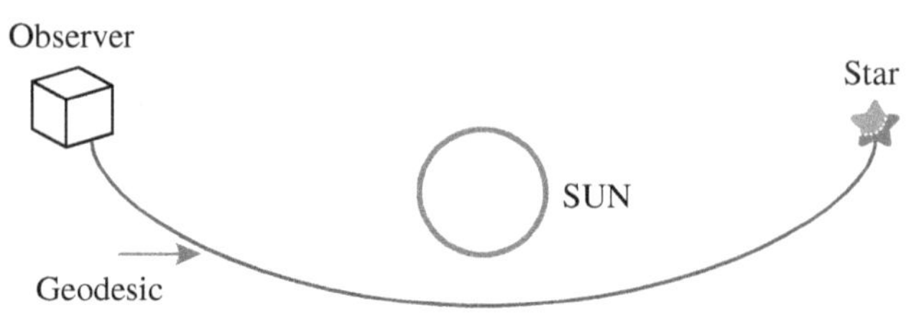

Fig. 3.1. Light from the star propagates along a geodesic path in curved space–time.

The curvature of space is what mathematicians study as geometry. Thus, Einstein transformed gravity into geometry — a mind-blowing and fascinating concept, indeed. The success of Einstein's theory also raised the possibility that other forces may have their origin in the geometry of space–time. The extra-dimensional theories discussed in Chapter 1 answered this in the affirmative. String theories made it much more appealing and mathematically rigorous, offering a promising route toward a unified description of gravity and quantum mechanics.

Another jewel in the crown of Einstein's theory of gravity was the recent discovery of gravitational waves from collapsing star systems. Gravitational waves are propagating disturbances in space–time produced by accelerating massive objects. The possibility of such waves was predicted by Oliver Heaviside in 1893 and later by Henri Poincaré in 1905, prior to Einstein's formulation of general relativity. They used the analogy between gravity and electric force to make this prediction. Just as an accelerating electric charge produces electromagnetic radiation, an accelerating mass produces gravitational waves. Once Einstein formulated the general theory of relativity, it became clear that his geometrical theory of gravity led to the existence of gravitational waves as "ripples in space–time." He never expected that the waves would actually be detected in the laboratory, since the force of gravity was extremely weak and detectable waves would require enormous masses undergoing intense vibrations to produce signals of measurable strength. It is therefore a testament to the ingenuity and perseverance of gravitational wave physicists that these waves were finally detected in 2015, exactly 100 years after Einstein proposed his theory. Gravitational waves "shake" a massive object on which it is incident, and these vibrations can be observed through the reflection of light waves on them.

The first person to search for gravitational waves was Joseph Weber of the University of Maryland. Although he did not succeed in detecting gravitational waves, his pioneering work provided the foundation for future monumental efforts, such as the Laser Interferometer Gravitational-Wave Observatory (LIGO) experiments in Hanford, Washington, and Livingston, Louisiana, which ultimately achieved the first direct detection. Prior to the groundbreaking 2015 discovery, indirect evidence for gravitational waves came from the observation of a binary pulsar by Russel Hulse and Joseph Taylor in 1974. The Nobel Prize related to gravitational wave discovery was awarded in 2017 to Kip Thorne, Rainer Weiss, and Barry Barish. There are now several other experiments in the planning stage in a worldwide effort to detect and study gravitational waves in greater detail, including VIRGO Collaboration in Europe, KAGRA Observatory in Japan, LIGO-India in India, and Geo-600 in Germany.

The observed gravitational wave signals in 2015 originated from the merger of two objects, which are remnants of stars, orbiting each other in a binary system. These objects can be either two black holes or a neutron star and a black hole. In general, when two massive astrophysical objects go around each other, gravity pulls them closer until they eventually collide and collapse to a single object. In this process, they emit gravitational waves, which are detected on Earth when massive test objects are shaken by the incident waves. So far, about 300 gravitational wave signals from different sources have been discovered. The Earth going around the Sun is also such a system that is constantly emitting gravitational waves, but they are too weak to be detectable. In the process of emitting gravitational waves, Earth is constantly losing energy and ever so slowly spiraling toward the Sun.

The force particle whose exchange gives the gravitational force is called a graviton, and the gravitational wave can be viewed as a manifestation of many gravitons.

3.4. Electricity and Magnetism

The next force that came to be understood in physics after gravity is that of electricity and magnetism. The existence of electricity was first noticed by ancient Greek philosophers, who observed that rubbing certain materials, such as fur and amber or a comb through hair, produces a force of attraction between them. One way to understand this is to assume that, in the process of rubbing, "something" is transferred from one object to the other, giving rise to a force. People initially believed that "this something" was a kind of fluid. It turned out not to be a fluid. The correct interpretation was eventually reached that all matter contains a property called electric charge, which is transferred during the rubbing process. From more detailed observations, it was concluded that electric charges are of two kinds: positive and negative. Like charges repel each other, and unlike charges attract. Before rubbing the two objects together, each contained an equal amount of positive and negative charges and hence did not exhibit any attraction since the positive and negative charges canceled each other. However, the process of rubbing transferred particles with negative charges from one object to the other, and both objects became oppositely charged and began attracting each other. Negative charges are transferred instead of positive charges because electrons carry negative charges and are considerably lighter than protons, which carry the positive charge. Using two of the same kind of "rubbed" objects, one can also establish that like charges repel each other. This led the French physicist Charles-Augustin de Coulomb to perform careful quantitative experiments and formulate the first precise law of electric forces, now known as Coulomb's law. Coulomb (1736–1806) was a French physicist and military engineer who devoted himself to scientific research when the beginning of the French Revolution forced him to move to a small village in France. He published his famous law of electric forces between 1785 and 1789. According to this law, like charges repel and unlike charges attract, and the strength of the force between any two charges decreases as the distance between them increases. This is similar to the way the law of gravity works, except that the gravitational force between masses is always attractive.

The next force intimately connected to electricity is magnetism. The Greeks suspected its existence from the strange property of a material known as lodestone, discovered around 600 BC. The Greek philosopher Thales of Miletus was among the first to study it, noting that it could attract iron without physical contact. These were naturally magnetized pieces of magnetite. The word "magnet" may have come from lodestones found in Magnesia, Anatolia, in northern Greece. Thus, magnetism came to be recognized as a distinct phenomenon.

Just like the electric force originates from electric charges, the magnetic force originates from magnetic poles: the north and south poles. Like electric charges, like poles of two magnets repel each other, and unlike poles attract. There is, however, a major difference between electric and magnetic forces: positive and negative electric charges can exist in isolation, whereas the north and south poles of a magnet always occur together. In a typical magnet, one end carries the north pole, whereas the other end carries the south pole. Nobody has seen a magnetic pole in isolation. An isolated magnetic pole is called a monopole. It is theoretically possible that magnetic monopoles exist, and several searches are underway to discover them.

It was discovered in 1820 by Hans Christian Ørsted that magnetism is related to electric charges in constant motion inside material objects. Ørsted observed that when charges move, they produce a magnetic field. We know that the Earth behaves like a giant magnet, which can align a compass in the north–south direction. This magnetic field is believed to have existed for the past 3.5 billion years and is caused by the electric currents of the molten iron core deep within the Earth. Thus, stationary charges give rise to electric forces, while moving charges produce magnetic effects, providing a unified description of two apparently different phenomena.

Michael Faraday extended Ørsted's work and found that when a magnet is moved toward a coil of wire, an electric current is induced in the coil. This showed that a changing magnetic field induces an electric current, further demonstrating the deep connection between electricity and magnetism.

The final important discovery in this field was made by Scottish physicist James Clerk Maxwell in 1865, when he published his paper "A Dynamical Theory of the Electromagnetic Field." His theory consists of four equations and is now considered to be one of the fundamental laws of physics. It embodied all the laws governing electricity and magnetism known at the time, including his own important modifications. Soon afterward, Maxwell realized that, when his laws are mathematically manipulated, they predicted the existence of electromagnetic waves that propagate through empty space at the speed of light — a result that revolutionized civilization and continues to do so in many ways. Wireless signals, a fixture of our daily lives, owe their origin to these waves.

3.5. The Weak Force

For a long time, scientists tried to understand all observed phenomena based on the assumption of only two known forces: gravity and electromagnetism. This reflected a simpler period when the internal structure of matter was not studied very much and therefore not yet well understood. As the atomic picture of matter gradually developed, the forces operating within atoms began to come into focus. This process began in the late 19th century, when evidence for an additional force appeared. As experimental and theoretical studies progressed through the first half of the 20th century, more properties of this force were uncovered. This force is called the weak force, so named because its strength was found to be much weaker than that of the electric force.

The first indication of the weak force appeared in 1896, when Henry Becquerel observed the phenomenon of radioactivity while studying the properties of uranium salts. He was continuing a line of research begun by his father, a Professor of Physics in Paris. Wilhelm Conrad Röntgen had just discovered X-rays, and Becquerel was investigating the nature of X-rays emanating from uranium salts when exposed to sunlight. During the experiments, he found that uranium salts emitted radiation even in the absence of sunlight. He studied further to check whether the emitted rays were X-rays and found that when placed in a magnetic field, these rays were being deflected. Since X-rays are not deflected by magnetic fields, he concluded that a new kind of rays came out of the uranium salt, and these rays had to carry an electric charge because something that does not carry an electric charge, like X-rays, will not be deflected by a magnet. The new rays were later established to consist of electrons and were the result of a new kind of force, the weak force, that operated inside the nucleus of the atoms. Becquerel shared the Nobel Prize for this discovery in 1903.

The weak force has a very short range, which explains why it does not manifest itself outside the nucleus and why it remained undiscovered for so long. It appears only when some nuclei are left on their own. It took several decades after its discovery to understand its true nature. Ernest Rutherford coined the term "beta rays" for the radiation emitted from the uranium nuclei, and the nuclear decay responsible for beta rays came to be known as *beta decay*. Subsequently, many nuclei were found to undergo beta decay, in which a parent nucleus transforms into another nucleus while emitting beta radiation.

We now understand that the weak force governs processes in which subatomic particles transform into one another. For example, a neutron can decay into a proton, an electron, and an anti-particle of the neutrino.

In the second decade of the 20th century, a major crisis arose in the study of beta decays. For many years after Becquerel's discovery, scientists did not know whether anything was accompanying the electron in beta decay or whether it was coming out alone. Around 1914, James Chadwick, Lise Meitner, Charles Drummond Ellis, and Otto von Baeyer found that in the radioactive beta decay of elements such as nitrogen and lithium, the electrons emitted from the nucleus did not have a fixed energy, as would be expected from energy conservation if an electron were emitted alone. Instead, it exhibited a continuous spectrum of energies. According to the law of energy conservation, in a physical process, the total energy at the beginning is the same as the total energy at the end. This would imply that electrons should be emitted only with a definite energy, since in beta decay, one nucleus transforms into another nucleus, whose energy is fixed. The electron's energy would then be determined by the difference in mass between the initial and final nuclei, according to Einstein's mass–energy relation. Since this was observed not to be the case, it led to a crisis in understanding this newly discovered process. Initially, it was suggested that energy might not be conserved in beta decays. This would have been a major departure from the established rules of physics, since energy conservation was, and still is, a cornerstone of physics that holds in all known situations.

This puzzle continued to haunt physics for more than a decade until Austrian physicist Wolfgang Pauli, in 1930, proposed a resolution by postulating that, in beta decays, an electron is always accompanied by a new particle called a neutrino. Since neutrinos and electrons can share the energy released in beta decay in different proportions, this explains why the emitted electron does not have a fixed energy. In the beginning, it was believed, even by Pauli, that the neutrino interaction forces would be too weak, making its detection in laboratory experiments almost impossible. However, evidence for the neutrino was finally obtained in 1956 by two American physicists, Frederick Reines and Clyde Cowan, in an experiment carried out at the Savannah River nuclear reactor. The theory of beta decays was developed by many scientists, following the ground-breaking work of Enrico Fermi and we now have a comprehensive framework known as the standard model, which describes the weak force along with all known forces and matter. We will discuss this in the following.

Weak interactions play a crucial role not only in the sustenance of life on Earth by enabling the Sun to emit solar energy continuously but also in the formation of elements in the periodic table that are fundamental to human life. In a typical weak interaction reaction, a proton and an electron appear when a neutrino strikes a neutron. The reverse process, where a neutrino and a neutron are emitted, occurs when an electron hits the proton in a dense environment of the Sun, thereby initiating the process of energy emission from the Sun.

3.5.1. *Age of the Earth and the Sun*

The weak force also helped resolve one of the brewing controversies of the late 19th century concerning the age of the Earth and the Sun. In the mid-19th century, geologists such as Charles Lyell, studying the Earth's crust, concluded that it formed through countless small changes occurring over vast periods of time. He proposed the idea of "uniformitarianism," which essentially boiled down to the statement that the age of the Earth was infinitely long. This view likely influenced biologists such as Charles Darwin, who proposed the theory of evolution that also required long timescales. However, Lord Kelvin disagreed. He applied the laws of physics known at the time to estimate the ages of the Sun and the Earth in various ways. Kelvin estimated the age of the Earth by calculating how long it took the Earth to cool from an initially molten state to its present solid form. Similarly, he argued that sunlight, the energy produced by the Sun, came from its gravitational shrinking. Using these methods, he estimated that the ages of the Sun and the Earth were between about 20 and 100 million years. This was in sharp contrast to the geological and biological estimates by Lyell and Darwin, leading to a significant scientific controversy. The resolution of this conflict came with the discovery of radioactivity, which made a huge difference to Kelvin's argument, since it provided new sources of energy at the core of the Earth from radioactive elements at its center, significantly slowing the cooling process and extending Earth's thermal lifetime. Similarly, at the core of the Sun, weak forces generate fusion energy that counterbalances gravitational contraction, allowing the Sun to shine steadily for billions of years. This understanding leads to the modern estimate that the age of the Sun is about 4.6 billion years, which is consistent with geological and biological evidence. The Sun is expected to continue shining for another 5 billion years.

3.5.2. *Mirror Asymmetry of the Weak Force*

As time went on and more was learned about the weak force, it was found to have a very unusual property. This property concerns how the weak force behaves under mirror reflection. Unlike the nuclear and electromagnetic forces that look the same (or are symmetric) under mirror reflection, the weak force does not exhibit this symmetry — a fact that puzzled many physicists in the 1950s. This feature will be central to the fundamental theme of this book — the concept of the mirror universe. Therefore, we briefly digress to explain what symmetry is, and in particular, what mirror symmetry means.

3.5.3. *Symmetry*

Symmetry exists when an object looks the same from different perspectives. A simple example is a car wheel. If it is mounted on a horizontal shaft and any markings are ignored, it appears identical when viewed from different angles. This implies that the car wheel has symmetry — specifically, circular (rotational) symmetry.

A long, straight road in an empty desert looks the same, no matter where you stand along it; if you move a small distance in either direction, its appearance does not change. This is called translational symmetry. When the shape of an object remains unchanged and looks the same, as the object is turned sideways or upside down, the object is said to possess symmetry.

Nature abounds in many kinds of such symmetries, some simple and some complex. Symmetry is a common feature in both the living world (plants and animals) and the inanimate world (planets, rocks, and the sky).

Manifest symmetries are always appealing. Starting with flowers, which are often radially symmetric and radiate beauty, to leaves, which often exhibit bilateral symmetry, and humans and animals, which are also bilaterally symmetric, symmetry is a mark of inner beauty and charisma. In the inanimate world, we see mountains with approximately conical shapes and crystals with various symmetric structures, among many other examples.

The history of science is full of examples where symmetries have played an increasingly important role in advancing our understanding. They have repeatedly helped broaden the frontiers of human knowledge. The concept of symmetry is so pervasive that mathematicians have devoted extensive efforts to studying it and uncovering its many useful properties.

Simple phenomena such as magnetism in metals, crystallization of compounds, and the melting of solids are governed by underlying symmetries. A substance is magnetic when it has tiny magnets inside, aligned in a row, so that all points along the alignment look the same — this is the translational symmetry of the tiny magnets. When a magnet is heated, the alignment of these tiny magnets is disturbed, as some of them flip direction due to the applied heat. This translational symmetry is lost, and the magnet then loses its attracting power (or magnetism). A ferromagnet loses its magnetism when heated above 770°C or 1417°F. The loss of symmetry causes a change in physical behavior. This simple example of symmetry breaking through heating has provided deep insights not only into magnetism but also into many other complex physical phenomena.

Coming back to the weak force, the particular symmetry relevant to its understanding is that of mirror reflection. In the first half of the

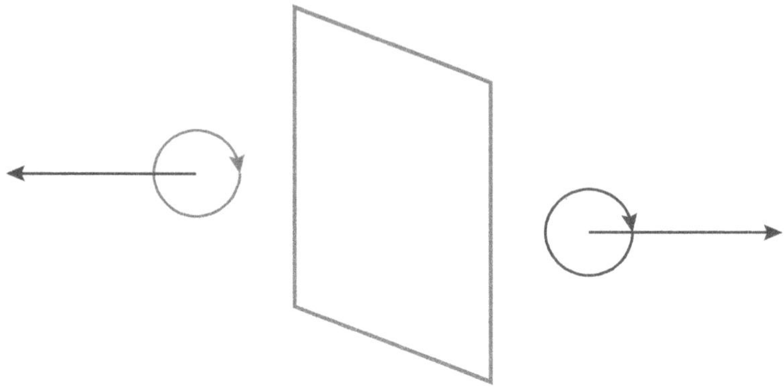

Fig. 3.2. The right-hand side shows the mirror reflection of the process on the left-hand side, where a spinning particle emits another particle in the direction of the arrow. If the force causing the emission is mirror symmetric, both sides will be observed experimentally with equal strength.

20th century, it was believed that all fundamental forces were symmetric under mirror reflection, including the weak force. Thus, when Chen-Ning Yang and Tsung-Dao Lee proposed in 1956 that weak forces do not respect mirror symmetry to resolve a certain puzzle in particle physics, it came as a profound shock to many physicists, who could not accept that the weak force had this odd behavior.

Figure 3.2 illustrates the mirror reflection of a spinning particle (with spin represented by a circle) undergoing weak decay, emitting an electron going along the direction indicated by an arrow. The mirror is represented by the central plane. Mirror invariance would imply that both the original process and its mirror-reflected counterpart occur with equal probability. However, in actual practice, one of the possibilities is not realized if it is a weak decay. This striking effect was first experimentally demonstrated in 1956 by Chien-Shiung Wu and her collaborators Ernest Ambler, Raymond W. Hayward, Dale D. Hoppes, and Ralph P. Hudson. They studied the beta decay of the cobalt nuclei, which decay into nickel while emitting an electron and an anti-neutrino. Lee and Yang were awarded the Nobel Prize for suggesting the groundbreaking idea behind this experiment.

3.6. The Strong Force

Recall that inside the atomic nucleus, protons and neutrons are bound together by the strong (nuclear) force. This nuclear force must have a very short range, since it is not observed outside the nucleus; otherwise, it would pull any nearby proton or neutron into the nucleus, thereby making the nuclei arbitrarily large. Scientists later established that this force acts on both protons and neutrons equally strongly.

What is the origin of the strong force? The first step toward solving this mystery was taken in a prescient paper by Hideki Yukawa, who suggested that the nuclear force arises from the exchange of a particle called a pion, as stated earlier. The pion couples strongly to protons and neutrons, giving rise to a strong attractive force between them. Yukawa's work introduced two profound ideas. First, it proposed that forces between particles can arise through the exchange of particles with integer spin (Fig. 3.3). Second, it showed the concept that the range of the force is determined by the mass of the exchange particle, as stated earlier. The heavier the mass, the shorter the range of the force. Yukawa published his theory in 1935. Although it was initially overlooked, strong support for his idea came with the discovery of the pion in cosmic rays in 1947. Yukawa was subsequently awarded the Nobel Prize in 1949 for this pioneering work. Today, there is a more fundamental theory of the strong force, where the exchange particle is called the gluon. This theory is known as quantum chromodynamics (QCD). David J. Gross, Frank Wilczek, and H. David Politzer were awarded the Nobel Prize in 2004 for their discovery of *asymptotic freedom*, a key property of QCD that explains how the strong force behaves at very short distances.

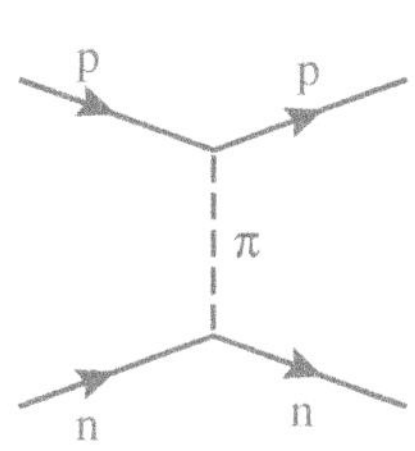

Fig. 3.3. Pion exchange between a proton and a neutron leads to the nuclear force; the finite mass of the pion limits the range of this interaction.

3.6.1. *Why the Proton and Neutron Have Nearly the Same Mass: Isospin Symmetry*

Earlier, we alluded to the mystery of why the two main constituents of the atomic nucleus, the proton and the neutron, have nearly the same mass, even though they carry different electric charges. We noted Heisenberg's suggestion that isospin symmetry is responsible for this near equality of masses.

To elaborate on this further, consider the following analogy. Draw two straight lines on a piece of paper with equal sides, arranged perpendicular to each other (Fig. 3.4). In your mind, label one line as the proton (p) and the other as the neutron (n). If the paper is rotated by 90 degrees, the proton line and the neutron line would be interchanged, however the overall picture of the two lines would look the same. It would not be possible to tell which line represents the proton and which represents the neutron. Figure 3.4 shows a simple, visual illustration of isospin symmetry between the proton and the neutron.

Heisenberg gave this symmetry the name "isospin," since similar mathematical properties were known in quantum mechanics for spin directions. However, he recognized that isospin is not a "real spin"; we do not see any real lines drawn in a nucleus, and the symmetry exists only in an abstract space. What is remarkable is that an imaginary picture like this would imply the near equality of the proton and neutron masses, which is measurable in the laboratory. The proton and the neutron are therefore said to belong to the same isospin family. Soon, Heisenberg's idea received support from the discovery of other groups of particles

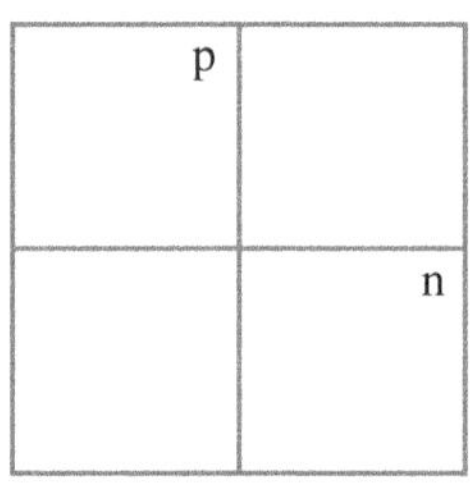

Fig. 3.4. Illustration of isospin symmetry responsible for the nearly equal masses of the proton and neutron.

with nearly equal masses, indicating that they too belong to other isospin families. Examples include the three pions (π^+, π^0, π^-), which have nearly identical masses, as well as the kaons (K^+, K^0), which also form the isospin-related states.

The success of isospin symmetry led physicists to consider more general types of symmetries. Imagine many identical pages with drawings similar to that in Fig. 3.4, distributed throughout space. If a symmetry could rotate all the drawings at the same time, regardless of their position in space, this is called a global symmetry. Alternatively, if each drawing could be rotated independently at different space–time points while preserving the symmetry, it would be called a local symmetry. Another example of a local symmetry, involving identical wheels on a common axle, is illustrated in Fig. 3.5.

Imagine a set of identical wheels mounted on a long, straight axle. The wheels are all free to rotate independently. The system has a particular appearance at any given instant. Note the arrow on one of the wheels. It represents a rotation of only that wheel, while leaving the others unaffected. Clearly, after rotation, the system looks the same. The same is true if any of the other wheels is rotated independently. This illustrates the concept of local invariance of a system. Since this invariance is maintained under any local rotation, or rotation of a single wheel, it is called a local symmetry of the entire system of wheels and axle. The same principle applies to physical systems when they possess a local symmetry. The local rotations may be more complicated than the simple mechanical

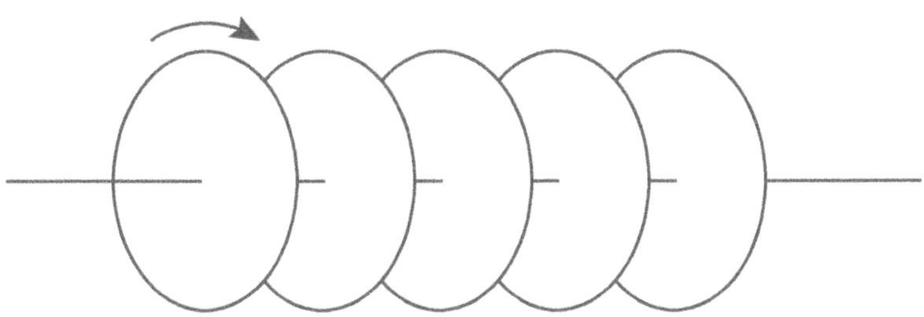

Fig. 3.5. Schematic illustration of local symmetry using identical wheels on a common axle. A rotation applied independently to any one wheel (indicated by the arrow) leaves the overall appearance of the system unchanged, illustrating local invariance.

rotation depicted here, but nonetheless, the underlying idea remains the same. More complex rotations correspond to more sophisticated symmetries.

3.7. The Birth of the Standard Model

The possibility of a single idea explaining all observations in nature is very attractive and has long fascinated scientists across many disciplines, and physicists are no exception. Einstein spent the last 30 years of his life trying to find a single, unified theory of nature — one law and one idea that would explain all phenomena in nature. His dream remained unfulfilled, but he helped place the search for unification at the forefront of physics. This quest has since inspired, and continues to inspire, generations of physicists to pursue that goal. The standard model represents a major, though incomplete, milestone in this ongoing journey toward a fully unified description of nature.

The basic idea underlying this pursuit has been the concept of local symmetries, as described in Fig. 3.5. This was first discussed by Hermann Weyl in 1915 and was generalized by Chen-Ning Yang and Robert Mills in 1953 to a form that proved far more useful for constructing modern theories, such as the standard model. At the time, Yang and Mills were working together at the Brookhaven National Laboratory in New York. Weyl's earlier work on local symmetries had already been applied to the theory of electromagnetic forces. Following the Yang–Mills formulation, physicists began to explore whether the weak force might also arise from a local symmetry. Julian Schwinger was among the first to apply the Yang–Mills framework in an attempt to construct a unified theory of electromagnetic and weak forces in 1960. Although his attempt was not successful, his student Sheldon Glashow extended his work in 1961 and took an important step toward the eventual development of the standard model.

The works of Schwinger and Glashow, however, still faced a significant problem: They predicted that weak forces were long-range, contrary to observations. This issue remained unresolved for several years until, in 1964, the idea of the Higgs boson was introduced.

3.7.1. *The Higgs Boson*

Local symmetries were attractive because they implied the existence of particles similar to the π-meson (Fig. 3.3), whose exchange can produce forces. These particles, called gauge bosons, were, however, different from the π-meson, since they carried spin one, whereas the π-meson had spin zero. The gauge bosons for the weak force were called W and Z. More importantly, the gauge bosons of Schwinger and Glashow had the problem that their exchange could only produce infinite-range forces. Therefore, without further change, they could not represent the weak force, since weak forces are short-ranged and confined to subatomic scales. The question then was how to create a short-range force out of a local symmetry. This is where the Higgs boson comes in.

The Higgs boson idea was proposed in 1964 by Peter Higgs, François Englert, Robert Brout, Gerry Guralnik, Carl R. Hagen, and Tom Kibble [22] in three separate publications.

According to their idea, the Higgs boson not only moves around as a particle but also has a part that is spread out and has a constant value over the entire universe. The Higgs boson is much like a pond of water. When the water is completely still, it is like the spread-out part of the Higgs boson, and the ripples on the surface of the water are like the Higgs particles that are produced in colliders. The spread-out part (hereinafter called the Higgs spread) is responsible for giving mass to gauge bosons like W and Z, as well as to all the standard model matter particles, such as quarks and leptons. Only those particles that interact with the Higgs field get mass from the Higgs spread. The stronger the coupling of the particle to the Higgs boson, the higher its mass, since the Higgs spread is the same for all particles. Thus, the significance of the Higgs boson is that it gives weight to all the particles that we know.

Steven Weinberg incorporated the Higgs boson into the model of Glashow and made it possible for local symmetries to describe short-range weak forces. The Weinberg paper came out in November 1967. At that time, its true significance was not realized. The author recalls a seminar by Tom Kibble (of Higgs mechanism fame) at Rochester and giving Weinberg's work a "lukewarm" endorsement, saying that "it was a rather interesting application of the Higgs mechanism." Little did anyone at the time know that this short four-page paper contained the correct code to describe all weak and electromagnetic forces of nature.

It took almost 45 years after the Weinberg paper for the Higgs boson to be discovered. It was the ripple part of the pond, or the Higgs boson particle, which was discovered in 2012 in two LHC experiments at CERN, Geneva, and the discovery was announced on July 4, 2012. It was the crowning achievement of the LHC, built under the beautiful mountains of France and Switzerland in the city of Geneva. More than 5,000 scientists from all over the world worked for several years to discover the particle. It was a monumental discovery. Peter Higgs of the UK and François Englert of Belgium received the 2013 Nobel Prize for the idea behind this discovery.

Of the force particles (the gauge bosons), the photon (γ) and gluons (G) are massless, since they do not couple to the Higgs boson. On the other hand, *W* and *Z* become massive and obtain their mass from the Higgs spread. The Higgs boson is the source of mass not only for the *W*- and *Z*-bosons but for all matter particles of the standard model. The only exception is the neutrino, which does not couple to the Higgs boson and is predicted in the standard model to be massless.

The *W*- and *Z*-bosons were discovered at CERN in 1983 by Carlo Rubbia and collaborators, and they confirmed the standard model. Many other effects predicted by the model have also been corroborated, making it an amazingly successful model of known forces. Everything is, however, not perfect with the standard model; it predicted that the neutrinos do not have mass, but neutrino mass was discovered in 1998, which means that the standard model is incomplete.

3.7.2. *Neutrino Mass*

The lightest of the standard model particles, the neutrino, has a complicated life story. Before 1930, physicists did not even know that such a particle existed. As discussed earlier, Wolfgang Pauli proposed the existence of this particle in 1930 to resolve a puzzle in beta decay observations, but many physicists were skeptical that this particle really existed. It was, however, finally discovered by Fred Reines and Clyde Cowan in an experiment conducted using the powerful fission reactor at the Savannah River Plant in South Carolina.

Today, neutrino research is a dominant theme in particle physics all over the world. Several laboratories are devoted to studying its properties exclusively, such as Fermilab in the US, J-PARC in Japan, and JUNO in China. This emphasis on neutrino research was not as strong a few years ago; so, why this shift?

After the neutrino was discovered, physicists thought for a long time that it had no mass and traveled at the speed of light, like the photon. In fact, it was argued that being massless was theoretically connected to the nature of weak forces being maximally mirror asymmetric. This offered a powerful theoretical argument to support this belief, which we now know was wrong. Experimentally, of course, it was hard to establish that neutrinos were massless. Only upper limits on their masses could be obtained, which were much smaller than the known masses of quarks, electrons, muons, and tau leptons. This made it clear that if neutrinos had mass, they must be very, very tiny.

The theoretical landscape in the early 1970s was focused on analyzing and understanding the standard model. Even though the standard model was almost fully constructed by 1970, Gerard 't Hooft's work established its mathematical consistency by 1973. In particular, his work showed how to systematically handle the infinities in the theory, thereby giving a complete picture of the standard model.

However, even as the final chapter of the standard model was being written, there were attempts to look beyond it in small incremental steps. One such step was the suggestion of the so-called left–right symmetric models, which tried to understand the origin of the mirror asymmetry of the weak force. In these theories, the observed weak interaction is a low-speed phenomenon [23] while the fundamental theory of weak forces is mirror symmetric, like other forces at very high energies. An immediate implication of the left–right symmetric theories was that the neutrinos must have mass. However, the theory did not say how large this mass must be, just as the standard model does not explain how large the different quark and lepton masses are. In particular, it did not explain why neutrino masses must be so tiny.

Things changed in the late seventies. During the period 1977–1979 period, a class of compelling theories was proposed that provided a framework for understanding why neutrinos might have extremely small masses. They were called seesaw models [24]. One of these papers, co-authored by the author with Goran Senjanović, was developed in the context of the left–right symmetric theories. This work was particularly influential because it connected the smallness of neutrino masses to the observed properties of weak interactions at low energies in a novel and compelling way.

On the experimental front, however, things were not so clear, and during 1977–1979, when the seesaw idea was first proposed, there was no evidence that neutrinos had mass. Experimental attempts were being made in different ways to see if they indeed have mass. A dramatic breakthrough occurred in 1998 at a biannual neutrino conference in Japan, where the Super-Kamiokande (Super-K) experiment announced that it had the first positive evidence for neutrino mass.

The conference room was electrified by this news because not only was the discovery revolutionary, but it was also the first major breakdown of the standard model. Newspapers around the world picked it up and announced this result to the public. Even the then President of the United States, Bill Clinton, mentioned this result in one of his speeches soon after. Curt Supplee, a *Washington Post* science writer, wrote on June 5, 1998, as follows: "A Neutrino Bombshell: It Has Mass."

How did the Super-K experiment, located under the hills of Mount Ikeno near the city of Gifu in Japan, conclude that neutrinos have mass?

The Super-K scientists were conducting a search for a phenomenon called neutrino oscillation using atmospheric and solar neutrinos. They used a huge vat of ultra-pure water surrounded by photomultipliers that detect flashes of light. If electron-type neutrinos passed through the water, they would produce a spark of light due to their interaction with molecules of water, and the photomultipliers in the water tank would pick up that light. Since the Sun emits neutrinos and there are neutrinos in cosmic rays, they expected to see neutrino-induced sparks inside the water volume as those neutrinos zoomed through the water. They indeed detected the sparks, but observed far fewer events than they expected and tried to understand the reason. They established that they are observing, for the first time, a new phenomenon called neutrino oscillations.

What is neutrino oscillation? In the standard model, there are three kinds of neutrinos: electron-, muon-, and tau-type neutrinos. In the 1960s, four scientists — Bruno Pontecorvo, an Italian, and Ziro Maki, Masami Nakagawa, and Shoichi Sakata from Japan — had independently predicted a phenomenon involving neutrinos called neutrino oscillation. It stated that one species of neutrinos in free flight can change its form and become another species if neutrinos have mass. Thus, if there were, say, 10 electron-type neutrinos flying from a source toward a detector, some of them could undergo neutrino oscillation and appear as another type of neutrino. If the device does not have the ability to detect the new neutrino that appears after oscillation, it will go undetected. In this situation, fewer electron-type neutrinos will be detected at the end of their flight [25].

In the 1960s, a similar oscillation phenomenon was predicted for the K-meson and its anti-particles by Murray Gell-Mann and Abraham Pais, who showed that a K-meson could transform into an anti-K-meson. This oscillation was already observed in experiments in the 1960s. The Super-K experiment was trying to detect the same kind of phenomenon for neutrinos using both neutrinos from the atmosphere and the Sun. Oscillations of neutrinos from the Sun were first observed by Raymond Davis Jr. in the 1960s. The Davis and Super-K observations established that neutrinos indeed have mass, falsifying the longstanding belief in massless neutrinos and demonstrating that the standard model is indeed incomplete. Arthur B. McDonald of Canada and Takaaki Kajita of Japan were awarded the Nobel Prize for the discovery of neutrino oscillations in 2015.

Chapter 4

Landscape of Darkness and Search for Dark Matter

"The answers you get depend on the questions you ask."

— Thomas Kuhn

Evidence for dark matter was becoming overwhelming toward the end of the 1970s. Fritz Zwicky's groundbreaking discovery was further supported by the works of Vera Rubin and Kent Ford. The urgency grew within the experimental community to unravel the cosmic mystery of what dark matter really is. If it is a cloud of particles, what particles constitute it? Or is it something else entirely? Researchers started exploring ways to search for it. How best to carry out such a search remained a major unresolved challenge that haunted physicists.

In physics, as in other fields of science, it is advantageous to have a rough idea of the target before embarking on a dedicated, expensive, and time-consuming endeavor. Experimentalists, therefore, turned to theorists for guidance. As we will see in the following, no particle within the standard model satisfies the requirements for dark matter. Luckily, this coincided with a period during which theorists had begun exploring physics beyond the standard model for other reasons.

Could these new physics scenarios provide viable candidates for dark matter?

Now that the standard model seemed to be working so well, theorists wanted to resolve its remaining conceptual shortcomings. The *W*- and *Z*-bosons, predicted by the standard model, had been discovered with masses consistent with theoretical predictions. They were discovered at the collider at CERN in Geneva in 1983. Several other predictions of the standard model were also confirmed by experiments. Despite its success, the standard model had issues that needed to be addressed. For example, why were the *W*- and *Z*-bosons not much heavier than their observed masses? The standard model did not shed any light on this question. Why was the force that distinguished matter from anti-matter so weak, as it was observed to be in 1964 in the decays of K-mesons? Why did neutrinos have mass? Scenarios of physics beyond the standard model were being proposed to address these issues. Some of them predicted interesting candidates that had just the right properties to be dark matter.

What is dark matter really like? Some basic facts about its nature were already known from astronomical observations. It must have no electric charge, otherwise it would reflect or emit light and would not be dark. It should not experience the strong nuclear force; otherwise, it would interact strongly with other atomic nuclei and bind to them, leading to the formation of heavier anomalous nuclei. Such anomalous heavy nuclei have been searched for by collecting samples from the deep ocean floor and analyzing them for the presence of elements beyond those already known, along with their heavier partners called isotopes. The ocean floor is the best place to look for them since, being heavier than water, they would have sunk to the bottom of the water column. To date, these searches have yielded no evidence for such anomalous nuclei, confirming that dark matter does not experience the familiar nuclear force.

History teaches us that once we have a theory for predicting a new particle, we can look for it using clues suggested by the theory and prove its existence or disprove the theory that predicted it. Such knowledge is also important in the search for new physics, as we seek to deepen and extend our understanding of nature. That is, after all, how the successful standard model was established as the theory of weak, strong, and electromagnetic forces.

The standard model describes all the properties of the familiar particles, such as protons, neutrons, and electrons, as well as the forces that act on them. It provides an almost complete picture of the universe at the present energies. Since none of the particles of the standard model can play the role of dark matter, its discovery would open a new path beyond the standard model, bringing with it new forces and new particles at higher energies. At the same time, it would not only broaden the frontier of our knowledge but also offer a deeper understanding of the universe.

In physics, simplicity of a theory and its experimental success often go hand in hand with the need for a broader, more unified, and conceptually satisfying framework. The discovery of such a framework has also historically led to new practical applications.

A classic example is the Biot–Savart–Ampère's law, which states that an electric current produces a magnetic field, and Faraday's law, which states that a changing magnetic field produces an electric field. These observations suggested that there is an intimate connection between electricity and magnetism. They were subsequently unified into a broader framework by James Clerk Maxwell in 1864, as noted in Chapter 3. He formulated a set of four equations that led to not only a unified understanding of electricity and magnetism but also the discovery of radio waves, light waves, and other forms of electromagnetic radiation, with enormous implications for human civilization. The true implications of the fact that electricity and magnetism were part of the same physics emerged only after the larger framework of Maxwell was conceived.

William Gladstone, who served as Prime Minister of England in 1868, once asked Michael Faraday, one of the great scientists of his time, what his novel research into electricity was good for. Faraday replied, *One day, Sir, you may tax it.* The same answer applies to much of present-day research into fundamental physics, including dark matter: Something whose value is not so obvious today may become enormously important tomorrow and even "a source of revenue!"

Other such examples include the study of quantum mechanics, which not only provided a more satisfactory framework for understanding subatomic physics but also enabled numerous practical applications, from lifesaving medical equipment to scanning devices in supermarket checkout counters to barcodes on your purchases, to name just a few.

The dark matter theorists have drawn inspiration from this historical pattern and looked to the landscape of physics theories beyond the standard model to find a new particle that satisfies the basic cosmological requirements for dark matter, and then searched for it. Once discovered, it could have its own applications. It could, of course, be that dark matter is completely outside the scenarios imagined so far in this chapter. That, too, would constitute a major discovery. This book discusses such a possibility.

Before we jump into searches for theories of new particles that play the role of dark matter, let us examine the particles already existing in the standard model and see if any of them can play this role. The only particles having some characteristics that allow them to qualify as dark matter (i.e., they have no electric charge and no nuclear force) are the neutrinos. However, dark matter plays an essential role in forming galaxies and other structures in the universe. Since neutrinos are extremely light particles, they are hot and energetic in the early stages of the universe's evolution. They are present in great abundance when structures (i.e., galaxies, stars, etc.) form. Due to their very tiny mass, they would be moving very fast and colliding with the structures that are forming. What their high speeds imply is that they would break up and erase any structure by their impact. This essentially disqualifies neutrinos from being candidates for dark matter. If neutrinos have cousins that are heavier by about 10,000–100,000 times, they will move more slowly and will not have the same destructive ability. They can then qualify as dark matter; we discuss them in the following. They are, of course, particles beyond the standard model and are part of new physics.

4.1. Super WIMPs: Time Running Out?

One direction of new physics beyond the standard model that has been the subject of extensive research since the 1980s is the supersymmetric theory. To understand what supersymmetry is, we need to give a few more examples of symmetry than those introduced in Chapter 3 in the discussion of weak forces and also examine the concept of symmetry breaking. Going back to the case of proton and neutron, we note that even though their masses are very nearly equal, there is a very small mass difference between them, roughly one part in a 1,000. To understand this, we must recognize the fact that symmetries in nature are often not exact and exhibit small amounts of symmetry breaking. In the case of the neutron and proton, isospin symmetry is broken by electromagnetic interactions. Thus, the concept of symmetry in physics is closely intertwined with that of symmetry breaking.

As an example of symmetry breaking, think of a straight rod standing exactly vertically on a perfectly level plane. An ant sitting on top of the rod will see all sides of it as identical. That is a symmetric situation. On the other hand, if you let loose the standing rod, it will fall to one side by itself after some time, and then the ant at the end of the rod will not see all sides around the fallen rod as identical. The symmetry is then lost from view. This falling down is an example of what physicists call spontaneous symmetry breaking. This is a concept that has also been used widely in constructing the standard model described in Chapter 3. The Higgs mechanism is an example of this. One can also think of symmetry breaking as arising from an external source. For example, in the case of the proton and neutron, the mass difference may be understood as arising from the electric charge of the proton. We will call this explicit symmetry breaking.

Supersymmetry is an extension of the idea of symmetry to a deeper level. It is a symmetry unlike those described above. It postulates that the two broad classes of particles in nature — matter particles (fermions, with half-integer spin) and force particles (bosons, with integer spin), introduced earlier in the book — may be related by a symmetry. This symmetry corresponds to an interchange between these two classes of particles that leaves the underlying physics unchanged. This states that matter and force particles would behave in very similar ways and would become indistinguishable in the symmetry limit. For example, it states that a particle like the proton, which is a matter particle, would have a supersymmetric partner particle, which is a force particle with spin zero, while light quanta, such as photons, which are force particles, would have a counterpart like a matter particle. The spin-zero supersymmetric partner of the proton would have the same mass as the proton if the world were exactly supersymmetric, since supersymmetry implies that there should be no way to distinguish between them except by their spin. They would also experience the same kind of forces acting on them with identical strengths. The matter and force particles are so different from each other that a symmetry relating them was inconceivable till the 1970s. When this idea emerged through the works of many people, it was named supersymmetry and quickly attracted widespread attention within the particle physics community.

No supersymmetric partner of the proton with the same mass as the proton has been found in any of the experiments carried out so far. In fact, there is no evidence for super-partners of any of the known particles in nature. Does this mean that supersymmetry is a wrong idea for particle physics? The answer is *no*. The absence of a same-mass super-partner is explained by the statement that supersymmetry, like many other symmetries in nature, is not an exact symmetry but a broken one, much like the fallen rod in the previous paragraph or the mass difference between the proton and neutron. Recall that the fallen rod originally came from a standing rod, which had the symmetry just like a proton super-partner with a different mass came from one supersymmetric theory. They both had the same mass before symmetry breaking. Symmetry breaking, much like the falling of the rod, can give the super-partner of the proton a different mass from the proton. That could have made it much heavier than the proton. If super-partner mass is pushed beyond the energy range of the Large Hadron Collider by a stronger breaking of the symmetry, this would explain why such particles have not yet been discovered. If one day a super-partner of any of the known particles is discovered, it would constitute a discovery of supersymmetry and a major breakthrough in our understanding of particle physics. The landscape of physics will be altered profoundly, opening new frontiers.

What do all these have to do with dark matter? To understand this, let us look at a supersymmetric version of the standard model. Such a theory contains spin-zero super-partners of the electron, neutrino, the various quarks, and spin-half super-partners of the photon, *W*- and *Z*-bosons, as well as the Higgs boson. In other words, every fermionic particle has a bosonic super-partner, and every bosonic particle (with integer spin, such as the *W*, *Z*, and photon) has a fermionic super-partner. If supersymmetry is broken, then the super-partners will be heavier, which explains why they have not been seen in current collider searches, as noted earlier. Of course, as the collider energies increase, these particles may eventually become accessible; however, if they remain undetected, supersymmetry will become a less compelling description of matter and forces.

To understand dark matter in the supersymmetric models, let us go slightly deeper into these theories. Consider the super-partner of an up quark, commonly referred to as an *up squark*. It would carry an electric charge of 2/3 and experience the strong force just like a quark due to supersymmetry. As a result, it does not satisfy the requirements for dark matter, since dark matter must be electrically neutral and should not have a strong force. On the other hand, the supersymmetric partners of the *Z*-boson or the Higgs boson, which are electrically neutral and do not have nuclear or electromagnetic forces acting on them, do qualify as candidates for dark matter. Moreover, since dark matter exists throughout the universe, the dark matter particle must be stable, i.e., it does not disintegrate into lighter particles over time. The super-partners of the *Z*-boson or the Higgs boson are collectively called "neutralinos" in particle physics. The lightest of these neutralinos is indeed stable since it cannot decay into any other particles due to the nature of supersymmetry. It also cannot decay into other neutralinos since they are all heavier than the lightest neutralino, which lacks the required energy to produce them. Thus, the lightest neutralino, often referred to as a supersymmetric weakly interacting massive particle (SUSY WIMP), is a well-motivated candidate for dark matter.

Since the SUSY WIMPs are heavy, the objections that ruled out the neutrino as a candidate for dark matter do not apply to them. Being much heavier than the electron or the proton, it would move very slowly in the current universe and would not be able to erase galactic structure. Particles in a colder gas move more slowly, and this is why a slow-moving WIMP is called cold dark matter (CDM). The lightest neutralino, or SUSY WIMP, is therefore an excellent candidate for dark matter.

This prospect generated considerable excitement among experimentalists. Supersymmetry also dictates precisely how the weak forces act on these super-partners, the neutralinos, the supposed dark matter candidates. This guidance has been crucial in designing experiments to detect them.

Also, following our earlier discussion, we can, within the supersymmetric framework, predict the present-day abundance of dark matter in the universe. When we discuss searches for dark matter, we will see that, despite massive efforts, these dark matter particles have so far eluded detection at levels expected from supersymmetry across much of their allowed mass range. This has led many scientists to conclude that supersymmetric particles are less likely to constitute the dominant component of dark matter.

4.2. Extra Dimensions and Kaluza-Klein Particle as Dark Matter

In our discussion of extra space dimensions in Chapter 1, we noted that Oskar Klein's interpretation of extra dimensions leads to towers of new particles when the extra dimensions are made compact (or curled up). The lightest of these new particles is stable and can therefore play the role of dark matter. This is a new kind of WIMP, not much unlike the SUSY WIMP discussed earlier. As in the SUSY WIMP case, the embedding of the standard model within the extra-dimensional framework determines the forces experienced by this dark matter candidate. This guides the experimentalists in devising their detection strategy.

4.2.1. *Physics and Cosmology of WIMP Dark Matter*

Even though SUSY WIMPS are now considered less likely to constitute dark matter, the idea that a generic particle with mass in the range of several tens of GeV remains a serious candidate for dark matter. However, it must satisfy certain key properties. A primary requirement is that its present-day abundance should match the observed dark matter density of roughly 10 millionth of the proton mass per unit volume. To achieve this, one must trace the evolution of the WIMP density as the universe expands. At some stage in this evolution, the excess WIMPs must annihilate with their own anti-particles into standard model particles in just the right proportion, so that the observed dark matter density remains today. This requirement, in turn, determines the interaction rate with quarks and leptons, which can then be used as a guide in designing experiments.

4.3. Neutrino's Heavier Cousins

The standard model discussed in Chapter 3 includes three neutrinos, which were originally predicted to be massless. However, the discovery of non-zero mass in neutrino oscillation experiments has added a new dimension to neutrino physics. It implies that there may exist new species of neutrinos beyond these standard model neutrinos with masses. The new neutrinos are called sterile neutrinos because they do not carry the weak force like the familiar neutrinos and are not produced in nuclear decay. The observation of neutrino oscillations indicates that the standard model is incomplete and needs extension to possibly include additional neutrinos that help generate neutrino masses. For a semi-popular summary of this story, see the book *The Neutrino Story* [25].

The sterile neutrinos are different from the supersymmetric partners discussed above, which arise if the original theory has supersymmetry. There is a class of theories known as left–right symmetric theories [23], where mirror reflection symmetry is a fundamental ingredient, and these theories predict sterile neutrino partners for the familiar neutrinos. The sterile partners of neutrinos have no electric charge and do not experience strong or conventional weak forces. They could therefore be ideal candidates for dark matter. We argued earlier in this chapter that the massive neutrinos of the standard model, which participate in neutrino oscillation to explain the solar and atmospheric neutrino deficit, do not have large enough mass to qualify as dark matter. However, if they had heavier partners, they could not move as fast and could stop playing the destructive role of erasing structure, thereby qualifying as dark matter candidates.

The "heavier" sterile neutrino, being much lighter than the proton, moves faster than the WIMP dark matter discussed above and is therefore called warm dark matter. As we will see later, in the discussion of the mirror world model, the warm dark matter neutrinos could also arise from a mirror sector of the universe.

The heavier neutrino cousins as dark matter also solve some of the problems of CDM (WIMP) models. For example, detailed calculations show that they do not predict too many satellite galaxies in the Milky Way as the simple CDM models do because it has just enough speed to wash out structure at the scale of smaller galaxies but not at larger scales, such as that of the Milky Way. They also address the core density problem, i.e., they do not lead to excessive accumulation of dark matter at the core of galaxies, unlike non-interacting CDM, again due to their higher velocities.

Typically, such a heavy neutrino is expected to decay with a very long lifetime into a lighter neutrino and a photon. This X-ray photon has been searched for in several experiments, such as the XMM-Newton satellite (an ESA mission with NASA participation). There were claims of observations of such X-ray signals from the center of the Milky Way galaxy as well as the Perseus cluster [26, 27], which is about 250 million light-years away from the Earth. This signal, however, has not been confirmed by subsequent searches. In any case, such an X-ray observation from galactic clusters in the future would provide a key test of the sterile neutrino dark matter model.

4.4. Skinny and Super-Skinny Axion

In modern particle theory, scientists sometimes propose new particles when doing so helps resolve a puzzling experimental finding. Of course, it should not disagree with any known experimental observations. This strategy has sometimes been very successful, leading the way to new areas of physics. The charm quark, whose existence was proposed by Sheldon Glashow, John Iliopoulos, and Luciano Maiani in 1969 before its eventual discovery, is an example of this approach. In earlier years of the 20th century, scientists were much more reluctant to suggest new particles; an example is the neutrino postulate by Wolfgang Pauli, who initially regretted introducing it simply because it was a new particle, even though it elegantly resolved a major puzzle of beta decay physics and was later confirmed experimentally, eventually becoming a central subject of modern physics.

The axion is such a particle that was invented by Roberto Peccei, Helen Quinn, Steven Weinberg, and Frank Wilczek to solve a major problem faced by the standard model. In the current theory of nuclear forces, known as quantum chromodynamics (QCD), there is an excessive strength for interaction that distinguishes between matter and anti-matter, i.e., a force that influences the behavior of particles in a way different from anti-particles. Such a force is called a CP-violating force. QCD has been extremely successful in describing the observations involving the nuclear forces. It predicted the above particle–anti-particle asymmetric forces that can have a variety of undesirable experimental effects, thereby posing a theoretical problem.

One tantalizing effect of such particle–anti-particle asymmetric forces in QCD is that they can cause a charge separation inside a neutron, leading to what is known as the electric dipole moment (EDM) of the neutron. If the positive and net negative charges are separated by a distance, the system possesses an electric dipole moment that can be measured by applying an electric field. An electric field will tilt a system with an EDM.

A neutron consists of one up quark, u, with charge 2/3 and two down quarks, d, with charge −1/3. Inside the neutron, they are arranged in a precise straight line, so that the net negative charge of the two down quarks and the positive charge of the up quarks fall on top of each other and cancel each other, making this dipole moment to be zero, as shown in Fig. 4.1. If there is any slight misalignment of either of the d-quarks with respect to this symmetric configuration, there will be a non-zero EDM of the neutron. The CP-violating QCD force alluded to above causes a large misalignment, thereby creating a serious theoretical puzzle.

Let us remind the reader that particle–anti-particle asymmetric forces must exist in the universe to explain the matter–anti-matter asymmetry

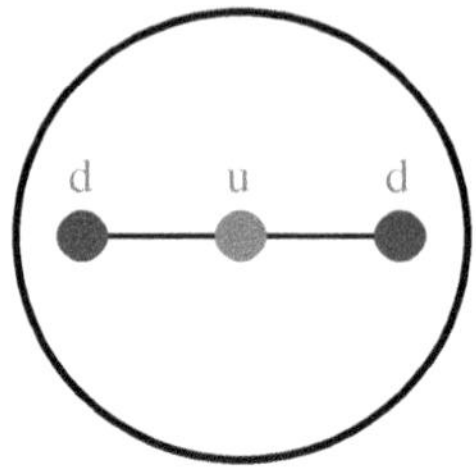

Fig. 4.1. Arrangement of quarks inside a neutron resulting in a zero EDM. Note the precise symmetric alignment of the charges.

that is the basis of life and civilization. Searches for such forces are therefore very important and constitute an important part of the particle physics research program. The neutron EDM, which results from this charge mis-alignment is one manifestation of this kind of effect, and searches for it have been carried out for many years. Even if non-zero, however, it is expected to be too tiny to account for the observed matter–anti-matter asymmetry. In any case, no positive result has been found so far. In fact, the sensitivity of these experiments has reached such a level that if a neutron were the size of the Earth, the positive and negative charges inside it would have to be misaligned by less than 1 micrometer (one millionth of a meter). Detection of such an exceedingly small charge separation inside a neutron remains a major experimental goal.

Taken at face value, matter–anti-matter asymmetric forces (CP-violating forces) arising from QCD can lead to a misalignment of quarks inside neutrons, which is about 10 billion times stronger than the upper limits set by current experiments. The standard model itself also contains a different source of such CP-violating forces, which has been experimentally established, but it predicts a much smaller misalignment of quarks inside the neutron (or a much smaller EDM for neutrons). One way to resolve this puzzle of QCD is to introduce a particle called the axion. The axion dynamically suppresses this QCD effect on the dipole moment of the neutron to zero. This proposal has generated significant interest in the particle physics community, and major experimental efforts are being devoted to its discovery. Although it has not been discovered yet, the idea behind this particle remains highly compelling.

What is the axion like? It turns out to be very different from any of the other dark matter particles considered so far. It is believed to be a superlight particle with mass, which is a trillion times smaller than the electron. However, unlike the electron, it is a bosonic particle. Its mass is not predicted by the theory and must be determined experimentally. It has the right properties to serve as the dark matter of the universe. Just as electrons are ubiquitous — present in every atom in our bodies and in all visible matter — if axions constitute dark matter, they will also be present everywhere. In fact, they will vastly outnumber electrons in terms of their cosmic abundance.

When Roberto Peccei and Helen Quinn [28] postulated the axion in the late 1970s, they argued that there is a new symmetry in the subatomic world that solves the above-mentioned neutron EDM problem of QCD. It was later shown by Steven Weinberg [29] and Frank Wilczek [30] that this symmetry leads to a new superlight particle, which came to be known as the axion. Subsequent studies demonstrated that for a certain strength of symmetry breaking, this superlight particle can fill the universe as a dark matter particle. Since it is ultra-light, it would exist in huge concentrations at any location in the universe.

An important property of this particle is that, if it impinges on a very strong magnetic field, it can generate a microwave beam whose frequency and strength encode the mass of the axion as well as how strongly it couples to light [31]. None of the other particles we know of behaves this way in a magnetic field. There are now dozens of experiments searching for axions using microwave cavities as well as other techniques.

No evidence has yet been found to support the existence of the axion. In principle, this particle can also coagulate in very large numbers to form what are called "axion stars" in the sky. A typical axion star could have a mass about 10 trillion times the mass of the Sun. They have been proposed as possible explanations for certain astronomical phenomena known as fast radio bursts, which are sudden, intense bursts of radio waves from the sky. Although prime sources of fast radio bursts are thought to be objects in the sky called magnetars, axion stars have also been considered as a possible source. The magnetars are made mostly out of neutrons, like neutron stars, but unlike neutron stars, the magnetic field inside a magnetar is 1,000 times stronger than that in a regular neutron star.

As is clear from the above discussion, the axion can decay into two photons. Dark matter axions can thus have important effects in astronomy by brightening up the sky through their decays into photons (light). There is some evidence for an excess cosmic light background observed by the New Horizons Spacecraft, which was launched in 2006 to study the planet, Pluto. After flying by Pluto, it has continued to observe deep space beyond and has reported an excess light from the sky. Although such experiments have a lot of background light to contend with and the observed excess could easily be light from distant galaxies and stars, it is something worth continuing to search for with as much sensitivity as possible, so that something definite about the connection of the excess light to axions could be established.

4.5. Dark and Hidden

Since we have not detected dark matter in laboratory experiments, but know of its existence from astronomical observations, one may wonder whether dark matter consists of particles that do not scatter off the familiar particles, such as protons, neutrons, or electrons. The detectors used to search for dark matter in the laboratory setups are made of protons and neutrons inside nuclei and electrons bound to them (see the following). If dark matter were of this hidden type, it is no surprise that laboratory experiments have failed to detect it. One can safely conclude that dark matter of this type will never be discovered this way. Other approaches must therefore be considered. Such particles could reside in a hidden world, yet still influence the universe through their gravitational effects. This hidden world will be called the mirror world, which will be discussed in Chapter 5. As we will see, this concept is distinct from the many worlds idea used to solve the "wave function collapse" problem in quantum mechanics, as discussed in Chapter 1. The mirror world is not fictitious but is believed to really exist, although it would remain hidden from us due to its very nature. It could, in principle, coexist alongside our visible universe. More about it later.

4.6. Light That We Cannot See

A photon is the carrier of light. Every beam of light looks continuous to us, but it is actually a bundle of light quanta, called photons. This suggestion was made by Albert Einstein to explain a new experimental observation, in which light incident on a certain kind of metal forces electrons to be emitted. This effect is called the photoelectric effect and is used nowadays to build solar cells and to generate solar energy. This suggestion brought Einstein the Nobel Prize in 1922.

How many bundles of light (or photons) there are in a beam of light determines how bright the light is. The color of light, on the other hand, is determined by the energy in each bundle of photons. When we see something with our eyes, it is because a beam of light or a collection of photons coming from the object excites our retina and produces an image. That is what enables us to see the distant stars in the night sky as well as a tree in our backyard or a bird flying in the sky. When electricity goes out on a stormy night, the house is dark because there are no photons anywhere in the house, and we cannot see anything. Electricity comes back, and so do the photons, and so does our ability to see things. The photons enable vision to mankind.

How are the familiar photons created? An atom, as we discussed before, consists of a nucleus containing protons and neutrons in its tiny core, with electrons occupying different energy levels, or orbitals, around it, following the rules of quantum mechanics. When an atom gets excited by some energy falling on it, the electrons in it gain energy and move to a different orbit of motion around the nucleus. When left to themselves, the electrons in the higher orbit get de-excited and fall back to their old orbits: The de-excitation process releases the extra energy that the atom had before the transition. That energy comes out as a photon, producing light. This process of de-excitation energy emitted as photons during electronic transitions was first proposed by Niels Bohr in 1913, as already discussed, and was so new and fundamental a concept that he was awarded the Nobel Prize for this work.

If a photon is the reason for brightness, what is a dark photon or light we cannot see? It sounds like an oxymoron to state that a particle responsible for light and vision could be "dark". Actually, the dark photon is not the real photon that we have been talking about. It is a name given to another kind of particle with properties very similar to a photon. It is a hypothetical particle present only in the theorist's imagination. Imagine that there is another world like ours with "people, trees, stars, etc." in it that reflect a different kind of light but are not sensitive to our kind of light. Then, we cannot see the "people, stars, and trees" of that world, but the "residents" of that world can see those things because their eyes are sensitive to that kind of light. Just as we use ordinary photons to see objects in our world, they would use dark photons to see objects in their world. Clearly, their photon is not visible in our world, and our photon is of no use to them. We call their photon the "dark photon" or light we cannot see.

This is somewhat like a bat, which hears a different frequency range of sound that we cannot hear, and vice versa. So, the bat's sound range is like "dark sound" to us.

Our photon is known to have no mass and always moves at the speed of light. If our photon had a mass, we could still see things using it, but only over limited distances. The smaller the photon mass, the farther you can see with it. Since we do not know how far the "residents" of the hidden world are able to see using their "dark photon," we do not know whether their photon has mass or not. Such dark photons themselves could also be the dark matter of our universe if they are massive enough. It could also be that they themselves are not dark matter, but instead act as a bridge between our world and the world of dark matter.

One way they could connect our world to the dark world is if the dark photon had the ability to convert into our photon. How rapidly it can do this is a matter of speculation, but it cannot be too strong because then all the dark world would be easily visible to us, and it should not be too weak if it is to be an effective connector to the world of dark matter. There are limits on how often the dark photon can turn itself into a real photon, from observations of supernovae in the sky (which occur at the end of a star's "life"). If a dark photon converted too efficiently into ordinary photons, it would drain a lot of energy away from a supernova explosion. However, from the observation of neutrinos from supernova 1987A, detected in 1987 by the Kamiokande experiment, we can conclude that all the energy from the supernova comes out as neutrinos and practically no energy is left to be drained by dark photon.

4.6.1. *Supernova: The End of a Star's Life*

We take a little digression to explain how a supernova happens. The starting point is a regular star that we see in the night sky. A star is a collection of a large number of atoms and nuclei within a definite volume of space in the universe. The stellar matter, consisting of these atoms, continuously exerts gravitational attraction on itself, just like any chunk of matter. Then why does the star not compress and collapse under this force? The reason is that at the core of a star, there are ongoing weak reactions every second that generate fusion energy continuously. This energy provides outward pressure in the form of radiation and thermal pressure, which pushes back against the collapse due to gravity. There is a subtle balance between the gravitational force pulling inward and the radiation pressure pushing outward that keeps the star stable. Part of the nuclear fusion energy comes out as radiation from the core, giving us starlight.

At the initial stage of the life of a star, energy is produced primarily through the fusion of four hydrogen nuclei into one helium nucleus, releasing two positrons, two neutrinos, and energy. Once all the hydrogen in a star is exhausted, the first phase of the star's life ends, and it becomes a helium-burning star. Then, helium undergoes a nuclear fusion reaction to form carbon that combines three helium nuclei. The resulting energy generated keeps the helium star in equilibrium against gravitational collapse. The carbon formed in the process undergoes fusion again, and the carbon nuclei fuse to form a silicon nucleus to do the balancing against gravity. This process continues till, at the "end," when an iron core is formed as the final product of successive nuclear burning stages. Iron has the property that it cannot release energy through fusion, and this is when there is no more energy being generated by the nuclear burning to push against gravitational collapse. The star then collapses under its own weight and undergoes rapid core collapse indefinitely. This collapse and compression process heats up the core and releases an enormous amount of energy. The released energy comes out primarily in the form of neutrinos, with only a small fraction emitted as electromagnetic radiation (light). The light emitted is so powerful in intensity that it produces a fiery explosion observed from Earth as a supernova. The small fraction of light creates a spectacular firework display in the sky. After the supernova explosion, what remains of the star is either a neutron star or a black hole, depending on how massive the star originally was. More massive stars collapse to black holes, while less massive ones form neutron stars.

When a supernova explosion takes place at the end of a star's "life," energy is released from the gravitational collapse, mostly as neutrinos. Normally, we would not detect these neutrinos due to their extremely weak interactions, even though they are produced in enormous numbers. Only the super-sensitive neutrino detectors on Earth can detect them. This is exactly what happened in 1987. The water detectors in Japan, known as the Kamiokande detector, which were built a few years earlier to search for proton decay, remarkably detected neutrinos from the supernova explosion of a massive star called Sanduleak, located in the Large Magellanic Cloud, about 168,000 light-years away from the Earth (i.e., light from this star takes 168,000 years to reach the Earth). The explosion actually occurred 168,000 years before 1987, and the Kamiokande detector happened to be perfectly timed to detect the neutrinos when they reached the Earth. This explosion is known as Supernova 1987A (SN1987A), named after the year it was observed on Earth.

Supernovae mostly emit anti-particles of the neutrinos, called anti-neutrinos. There were 12 anti-neutrinos observed on February 24, 1987, in the Kamiokande detector; the Baksan detector in Europe saw five, and eight were seen in the IMB detector in the United States, making a total of 25 anti-neutrino events. All events occurred at roughly the same time but had different energies, confirming that they came from the same source, SN1987A.

We were just lucky to have the Japanese detector Kamiokande as well as the other detectors operational about 5 years before this event, allowing us to detect these supernova neutrinos. If we were a few years late, we would have missed this momentous event and the wealth of knowledge that came with it. What an amazing coincidence!

The total energy emitted from this supernova could be estimated from the 25 anti-neutrino events observed, and it was found that it roughly matches the expected energy emitted from a supernova based on this solar burning theory described above.

Why were there so few anti-neutrinos seen on Earth if the SN1987A emitted all its energy in anti-neutrinos? The reason is that the original anti-neutrinos were emitted in all directions into a sphere surrounding the supernova, and Earth occupies only a tiny fraction of that sphere, so only a small number of them were detected. However, this small number contained enough information to estimate the total energy output of the supernova. If there were any new physics that would cause emission of particles, such as a profuse beam of dark photons from the supernova, it would take energy away from the neutrinos, reducing the number of neutrinos detected. Nevertheless, since all the energy produced in the process of supernova collapse is observed to be carried by the neutrinos, there is no room for the production of any other particles. This, therefore, puts a limit on the extent of dark photon–photon conversions. Many other such new physics scenarios were constrained by the observation of the 25 anti-neutrino events seen from SN1987A.

4.6.2. *Searching for a Dark Photon in a Collider*

It can happen that the dark photon transforms to the normal photon a very small fraction of the time, and in such cases, it can be emitted by electrons and other light electrically charged particles from our world.

The dark photon would then decay in a very short time interval to electrons and positrons. Within that very short time, it will travel a long distance because it moves very fast. This feature can be used as a technique for identifying dark photons. Normal photons do not decay. This technique has been used in various accelerators that have electron beams to search for the dark photon. Dark photons can also be produced in proton–proton collisions at the LHC, and such processes have also been exploited in these searches. However, no evidence has so far been found for the existence of dark photons.

4.7. Black Holes as Dark Matter

Unlike Newton's law of gravity, which states that gravity only attracts massive objects (or objects with weight), Einstein's theory of general relativity states that gravity also attracts light. This follows from Einstein's special theory of relativity, which states that mass and energy are equivalent (the famous $E = mc^2$ formula). Since a light beam has energy, it is affected by gravity in a manner analogous to mass, and a gravitational force attracts it. Thus, a sufficiently strong gravitational force can therefore pull a light beam hard and turn it around. So, for a big chunk of matter like a star from which light is coming out, as the star becomes increasingly massive, it can pull the light beam inward, and eventually, when all light is pulled back, the "star" will look dark. Such an object is called a black hole. Also, by the same token, any light that falls on a black hole never gets reflected since the reflected light will be pulled back. The chunk of matter remains, forever, dark. Since black holes cannot emit light and remain invisible, could they play the role of dark matter? To find out, we have to know a bit more about their physical properties.

Another way black holes can form is as follows: In the very early universe, when matter existed at extremely high temperatures, it was sloshing around very rapidly. At that epoch, the density of matter at some location becomes so large that it attracts all light back to itself, forming what is known as a "primordial black hole." This idea was suggested by Stephen Hawking in 1971. These black holes could either grow by absorbing matter around them, or they could just remain small. Their masses could be as small as a millionth of a gram or as large as 10,000 solar masses, according to an estimate by Stephen Hawking and his student Bernard Carr in 1974.

Stephen Hawking was an enormously inspiring scientist who, despite suffering from an incurable disease for almost all his life, continued to contribute many ideas in physics. He died in 2018 at the age of 76.

Since black holes are naturally dark (they do not emit light and absorb all light falling on them), they are, in some sense, natural candidates for dark matter.

So, could they be the dark matter? If they were, that would relieve the pressure on scientists to keep searching for new particle candidates, since black holes are made only of known matter, such as protons, neutrons, and electrons.

To understand whether black holes can be dark matter, we need to know a bit more about them. A very surprising property of black holes, pointed out by Stephen Hawking, is that black holes are not completely "black" as the name indicates; in other words, it is not that they do not absorb everything and don't emit any radiation. They indeed do emit radiation known as the Hawking radiation.

It is a bit confusing at first sight, since we just finished arguing that black holes attract everything. So, how can they radiate anything? It turns out that it is due to the property of quantum mechanics that something like Hawking radiation happens. A simple way to understand this is that near the surface of the black hole, a strong gravitational field can produce a particle–anti-particle pair by a quantum mechanical effect, with both particles and anti-particles traveling in opposite directions. The members of this pair that are traveling away from the black hole will then escape, and these escaping particles are observed as radiation emitted from the black hole. How hot the radiation is depends on how heavy the black hole is: The heavier the black hole, the lower the temperature of the radiation.

Hawking radiation is similar to the heat emitted by a hot charcoal and obeys the same kind of laws. They are known as blackbody radiation and follow a law given by Max Planck in 1900. As the black hole radiates energy, it will lose mass since the energy comes from the mass of the black hole. Recall that Einstein's theory states that "mass and energy are two manifestations of the same thing and can be converted to one another." As the black hole mass gets smaller, its temperature gets higher, and therefore it radiates more energy. Eventually, it disappears, although this process can take an extremely long time. For example, it has been estimated that a black hole as massive as the Sun (about 10^{33} grams) will take 10^{64} years to fully evaporate. This is much, much longer than the age of the universe, which is only 13.8 billion years. A black hole that was present in the early moments of the universe will just be decaying today if it had a mass of 10^{14} grams. So, certainly, black holes of much smaller sizes than the Sun could still be around and could very well be dark matter.

The next question is: If all the dark matter were made of black holes, where would so many black holes come from? We just saw that one way black holes form is from stars dying. Each time a star dies, there is a supernova explosion. If all dark matter black holes were to arise this way, we would need so many stars to die that it would contradict observations of the rate of supernovae. Another way black holes could form is during the early moments of the universe from radiation and matter sloshing around, as mentioned above. They form primordial black holes. We do not know how many primordial black holes there were to start with, since they don't produce supernova flashes. As a result, they could account for all of the dark matter without contradicting any observations [32, 33].

This idea of black holes (Fig. 4.2) being dark matter has had its ups and downs. However, it received renewed interest with the discovery of gravitational waves, detected by LIGO. These waves were determined to come from two black holes rotating around each other and collapsing. This discovery gave new life to the idea that black holes may be more abundant than previously thought and therefore could be dark matter.

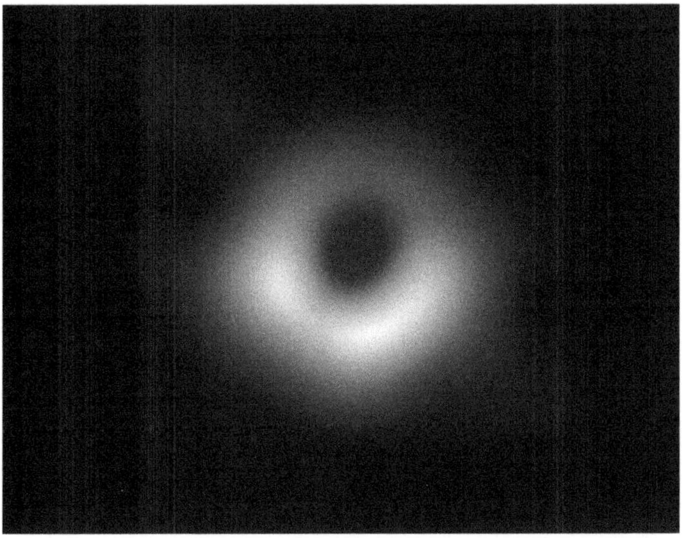

Fig. 4.2. Image of a supermassive black hole, showing its central dark shadow surrounded by a bright, asymmetric emission ring.
Source: ESO (CC-BY 4.0).

Various aspects of this possibility are currently under intense scrutiny by theorists. No firm conclusion can be drawn yet.

4.8. Milli-Charged Dark Matter

The list of possible dark matter candidates continues to grow. Imaginative theorists keep proposing more particles that could qualify as dark matter. It is important to explore as many possibilities as one can, since the true nature of dark matter remains unknown. Another class of particles that could also qualify as dark matter candidates are the so-called milli-charged particles.

Electric charge is a fundamental attribute of matter: for instance, in atoms, we encounter three types of particles: positively charged (protons), negatively charged (electrons), and uncharged or neutral (neutrons). When you examine all observed charged particles, there is a very interesting conclusion one can draw: All charged particles carry electric charges that are multiples of a fundamental positive or negative charge on the protons and electrons, e.g., a negative charge on any observed particle is some integer (1, 2, 3, 4,...) times the charge of the electron, and a positive charge similarly could be an integer times the charge of the proton. The electric charge of the electron was measured by Robert Millikan in 1909, for which he was awarded the Nobel Prize in 1923. This led physicists to conclude that there is a fundamental unit of electric charge, and no particle could have charges that are not as discussed above. This is called charge quantization, and understanding its origin remains a fundamental puzzle in theoretical physics. There are various arguments as to why electric charges should be quantized, but it is still an open question whether all charges are really quantized or whether there may be particles that have smaller charges than the electron or proton. These latter kinds of particles are collectively called milli-charged particles. There are various limits on how large such charges can be for a given mass, but we do not get into these technical details here. In certain theoretical frameworks involving two photons, a normal and a dark photon, which mix with each other, particles that couple to the massive dark photon appear as milli-charged [34].

It has been pointed out that milli-charged particles, if they exist in nature, can have such a small charge that they are not visible to ordinary light and so can escape detection in the experiments done so far. They will also be stable since all known particles have integral charge, and electric charge conservation will prevent the milli-charged particles from decaying into other known particles. Thus, milli-charged particles could easily serve as viable candidates for dark matter of the universe and have been entertained as such. A variety of experimental searches have been proposed in the literature to search for milli-charged dark matter.

4.8.1. *List Goes On*

So far, we have described dark matter as consisting of only particles. However, there is no fundamental reason to assume that dark matter is a bunch of particles spread out in the universe, although this idea provides a very convenient framework for both theoretical discussion and experimental searches. It is also possible that dark matter exists in other forms, such as extended objects, like dark rocks, dark planets, or other macroscopic structures. At this stage, our knowledge about dark matter is so limited that it is prudent to consider a wide range of possibilities. Exploring diverse scenarios helps ensure that viable possibilities are not overlooked. In fact, it has been conjectured that dark matter could be an extended, macroscopic rock-like object made out of invisible matter and not necessarily tiny point-like invisible particles floating all around. The main challenge is to construct a plausible theoretical framework for such an idea.

Within a certain class of particle models, it has been proposed that dark matter could form massive blobs of large rock-like mass, just floating around in space and spread throughout the universe, whose gravitational effects account for the observed phenomena. Since most of the evidence can be understood using the gravitational forces, there remains considerable freedom for such speculative ideas. The goal is to identify which, if any, of these possibilities is realized in nature.

How could dark matter form rock-like objects, which could be two blocks long? As an example of an underlying theory, consider a particle physics scenario where the particles have secret interactions that are attractive, strong forces, but no dark electric charges or any other secret repulsive force. In such a scenario, dark matter particles will keep attracting each other till they are so large that the particles at the edge are not attracted enough to be held together. The edge particles are then pushed out by the pressure of the rest of the matter in the rock. The object then has a size that is large and depends on how many particles coagulate together. Investigations have shown that this type of dark matter could easily form nuclear blobs with sizes as large as 100 meters and masses ranging from a few grams to a few hundred tons. For comparison, recall that ordinary matter composed of protons and neutrons cannot form such large nuclei, mainly because of the electric repulsion among the protons. Their sizes are typically of the order of a fermi or a trillionth of a centimeter [35].

Another class of extended dark matter objects could also appear in the form of large diffuse clouds. If such a cloud were to approach a black hole, there would be a big push and pull between the cloud of dark matter and the black hole. This will lead to the emission of gravity waves, which can be detected by existing gravity wave detectors, such as LIGO. Such searches have been attempted recently [35], although the result was negative. Nevertheless, these studies served to place important constraints on the possible size and properties of such dark matter clouds.

4.9. Searching for Dark Matter: Make It, Break It, Shake It, or Sniff It

Now that there is little doubt among astronomers and particle physicists that dark matter exists and that it is most likely an elementary particle, the question of its nature has become increasingly urgent. Of course, it could also consist of black holes that are made out of conventional matter; no new particles would be required, and the urgency of particle-based searches would be reduced.

If dark matter is a new particle, it may either have (i) very feeble interactions with the familiar matter, such as protons, neutrons, and electrons, or (ii) no interaction at all other than the gravitational force, which is a universal force acting on all matter. For the first class of particles (case (i)), the search strategy depends on the detailed nature of their interactions with familiar matter. The second class of dark matter particles (case ii) can possibly be probed only through astronomical observations, since their interactions with standard model particles, which make up detectors, are extremely feeble or entirely absent.

The third class of dark matter is the axion, which is an extremely light particle with very feeble interactions with familiar standard model matter but possesses interesting properties, as discussed earlier. Detection of such particles requires techniques different from those used in cases (i) and (ii). This type of dark matter is often called "wave-like" dark matter.

Broadly speaking, there are three major strategies to search for heavy dark matter. These are often described as "make it, break it, and shake it." The "make it" approach refers to producing dark matter particles in particle colliders and then detecting their presence through missing energy signatures or associated visible particles. The "break it" approach refers to the collision and annihilation of dark matter particles in space to produce familiar particles that can be detected using telescopes and other detection instruments. Finally, the "shake it" approach refers to the detection of naturally occurring dark matter particles in our galaxy by observing their collisions with normal matter, such as electrons or nuclei in a detector, where these interactions deposit small amounts of energy, producing detectable signals in the form of heat or light.

A fourth approach, called "sniff it" here, involves detecting faint background signals or subtle environmental effects induced by dark matter, although this terminology is less commonly used.

4.9.1. *Collider Searches ("Make It")*

This technique involves producing dark matter in particle collisions at an accelerator and studying its properties using familiar particle physics methods. An ideal collider for this purpose is the proton–proton colliding machine, such as the Large Hadron Collider at CERN, which can produce particles with masses up to a few thousand times that of the proton. Typically, dark matter is produced together with other familiar particles, such as quarks, electrons, or muons. Since the forces between the familiar particles and dark matter are very weak, dark matter particles, once produced, will escape from the detector without leaving a track. Recall that this is unlike an electron, which, as it passes through a detector, will tear off electrons from the detector atoms, leaving a bubble track or other sort of track in the detecting device.

This technique is known as the "missing energy" signature of dark matter because the energy carried away by dark matter leaves no direct trace in the detector. If dark matter is produced in a proton–proton collision, the total incident energy of two colliding protons before the collision will be higher than the total measured energy of all particles observed after the collision. This mismatch leaves a gap between the initial and final energy contents in the collision event, leading to the missing energy signal. Using the amount of "missing" energy, one can infer the mass scale and other properties of the dark matter candidate.

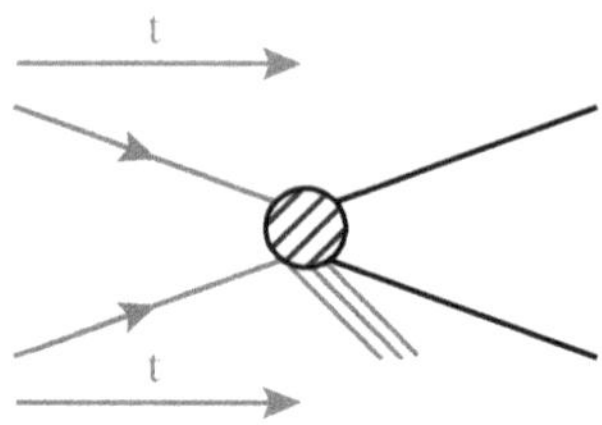

Fig. 4.3. Schematic illustration of a typical "make it" search. The incoming standard model particles are shown as lines converging to a collision point, where they produce dark sector particles in the final state, represented by solid lines emerging from the vertex. Additional standard model particles produced in the interaction are indicated by closely spaced lines. The arrows denote the direction of time.

So far, no trace of such missing energy signals has been discovered at the LHC. Since the mass of dark matter also affects its production rate in proton–proton collisions, experiments can place upper limits on the production cross-section for a given mass. As dark matter experiments are becoming increasingly sophisticated, the upper limits on the production rate keep decreasing. Even though the collider searches for dark matter have come up empty so far, they have provided valuable information regarding which kind of particles cannot be dark matter, thereby narrowing the theory landscape for such particles. In fact, the failure of such processes to find the supersymmetric dark matter particle (super-WIMPs) has made supersymmetric theories of dark matter less compelling. A schematic illustration of a typical "make it" search is shown in Fig. 4.3.

4.9.2. *Underground Detectors ("Shake It")*

Another class of dark matter search experiments relies on the fact that dark matter particles permeate our galaxy almost uniformly and move with typical speeds of about 300–400 kilometers per second. The exact speed, of course, depends on the mass of dark matter particles, with heavier ones moving more slowly than lighter ones.

The principle behind "shake it" dark matter searches is illustrated in Fig. 4.4.

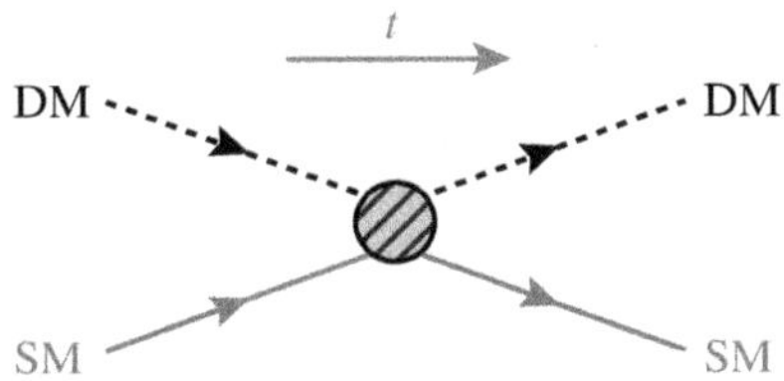

Fig. 4.4. Schematic illustration of the principle behind "shake it" dark matter searches. A dark matter (DM) particle scatters off a standard model (SM) particle in the detector, transferring energy to it. This interaction can produce a measurable signal, such as heat or the ejection of an electron from an atom in the detector material. The arrows indicate the direction of motion, and time progresses along the indicated direction.

This discussion of speeds applies to the class of WIMP dark matter described above. Note that dark matter must be in constant motion; if it were at rest, it would fall toward the center of the galaxy under gravitational attraction. This is much like the way that Earth is in constant motion around the Sun to keep going, and without this motion, it would fall toward the Sun due to the Sun's gravitational attraction and would burn out. It is the orbital motion of the Earth that balances the attraction of the Sun's gravity. Try whirling a stone tied to the end of a string. If the stone is moving rapidly, it will continue in circular motion, but as soon as it slows down, the string will slacken, and the stone will fall inward toward the hand.

When dark matter particles are in motion, they possess kinetic energy. If we put a detector on the Earth, some dark matter particles will collide with it and cause nuclei (or protons) in the detector to recoil, producing a measurable heat signal. Alternatively, a dark matter particle may scatter off an electron and eject it from an atom. This nuclear recoil or the tracks created by the ejected electron can be detected and used as a signal in dark matter searches.

We do not know the mass of dark matter particles. If the particles are much heavier than a proton, then they can bounce off the nucleus and lead to the recoil signal. On the other hand, if they are much lighter than a proton, the dark matter collision cannot push the nucleus to move, but it can easily knock off electrons from the target atom. These are referred to as heavy and light WIMPs and are the focus of many experimental searches. However, they cover only a small fraction of the full range of possible dark matter masses. The heavier and moderately light dark matter candidates are generally better motivated by particle physics models than many alternatives, except for axions, which are superlight and also have strong theory motivation, as discussed earlier.

If dark matter is lighter than an electron, it can be searched for using alternative techniques, such as the emission of sound waves (phonons) in a detector material.

Dark matter detectors are generally located underground to avoid fake signals coming from cosmic ray particles, which are also moving very fast and can collide with the detector material and mimic a dark matter signal. Once underground, dark matter particles, due to their extremely weak interactions with ground matter, can pass through the overburden and reach the detector in relatively higher numbers, whereas most cosmic ray particles are blocked from the detector, since they will not be able to penetrate the surface cover of the Earth above the detector.

There are various kinds of materials that can be used to detect dark matter. One of the most widely used is liquid xenon. Xenon is an inert gas present in trace amounts in the atmosphere. It can be captured and liquefied by cooling it in a bath of liquid nitrogen. Xenon becomes a liquid when cooled below -162°C, while nitrogen becomes a liquid below -196°C. Thus, because liquid nitrogen has a lower boiling point than xenon, any xenon gas in a bath of liquid nitrogen will condense into liquid xenon. This liquid xenon, when exposed to dark matter particles, can be ionized through collisions with massive dark matter particles. In the ionization process, the electrons are liberated from xenon atoms, carrying away their share of energy from the colliding dark matter. In this process, they will leave a track in the liquid as they transfer part of the dark matter's energy in the collision to the detector matter. Experiments that use this material include the following: the XENON1T experiment in Italy, the LUX and LZ experiments in the

US, the PandaX experiment in China, and the XMASS experiment in Japan.

The XENON1T experiment is located at the Laboratori Nazionali del Gran Sasso (LNGS) in Italy, between the towns of L'Aquila and Teramo, about 80 miles east of Rome, under about a mile of thick rock that provides protection to experiments from cosmic rays hitting the Earth.

The LUX-Zepelin (LZ) experiment is located in Lead, South Dakota, US, at the Sanford Underground Research Facility (SURF), deep beneath the Black Hills, with a rock cover comparable to that of XENON1T.

The PandaX experiment is located in the China Jing Ping Underground Laboratory in China. It is one of the deepest such experiments, located at a depth of one and a half miles.

The XMASS experiment is located in the Kamioka Observatory in Japan, beneath roughly two-thirds of a mile of rock.

The LZ, PandaX, and XENON1T experiments, along with other searches, have so far yielded negative results in the search, setting stringent limits on the scattering rate for various dark matter masses.

Other kinds of materials used for dark matter detection include semiconductors (such as silicon-based materials) and superconductors owing to their sensitivity in detecting collisions. Another material used for this purpose is germanium. An experiment called EDELWEISS was among the first to use germanium detectors for dark matter searches. The Super CDMS experiment, located 6,800 feet underground in a nickel mine in Canada, uses a combination of germanium and silicon as detector materials.

Another detector material used is sodium iodide. This is used in an Italian experiment known as DAMA, located in the same Gran Sasso underground facility as XENON1T. It looks for a seasonal variation of the signal from dark matter scattering off the detector material as the

Earth orbits the Sun. The relative velocity of dark matter with respect to the Earth changes as the Earth is moving one way versus another around the Sun, whereas the dark matter is moving only one way. Since the signal rate depends on the relative speed of the dark matter with respect to the Earth, there will be a periodic variation in the signal rate with a 1-year cycle. The DAMA and its upgraded version, DAMA-LIBRA, have reported a positive signal; however, this result has not yet been confirmed by other independent experiments.

The same kind of detectors can be used to detect dark blobs if they have weak forces connecting them to visible matter, such as neutrons. As they pass by the detectors, they can shake the detectors, generating heat that can be sensed by the device.

4.9.3. *Indirect Signals from the Sky ("Break It")*

There are several experiments in the sky looking for the weakly interacting massive particle (WIMP) dark matter (Fig. 4.5). One kind of detector on board the International Space Station is the Alpha Magnetic Spectrometer (AMS). Another is the Fermi Large Area Telescope (Fermi-LAT), which observes energetic gamma rays (or photons). The AMS detector, an international collaboration, was launched in 2011 aboard the Space Shuttle Endeavour and installed on the International Space Station, involving scientists from the US, France, Germany, Italy, Japan, and Sweden.

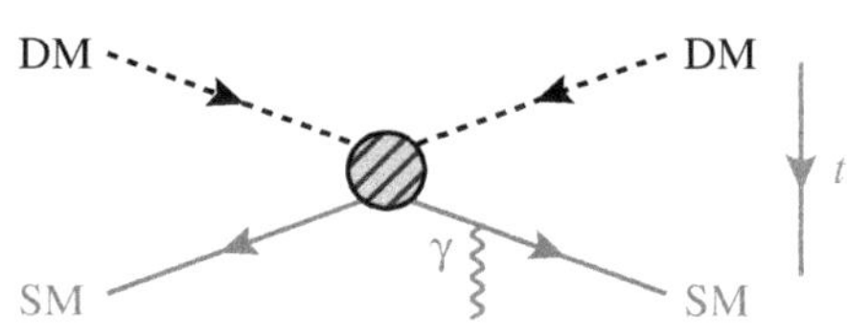

Fig. 4.5. Schematic illustration of "break it"-type dark matter detection. Two dark matter (DM) particles collide and annihilate to produce familiar particles, which subsequently emit photons. These signals can be observed by space-based or ground-based detectors. The arrows indicate particle trajectories, and the marked direction represents the flow of time.

Such detectors measure the flux of cosmic rays passing through them. There is some expectation from scientific analysis as to how many cosmic ray particles should be observed in the absence of dark matter annihilation. Any excess signal in the detector must therefore come from additional sources, including dark matter as one of those possible sources.

The Fermi-LAT experiment, in particular, has collected a large amount of gamma-ray data by observing the dwarf galaxies, which are believed to be richer in dark matter content compared to large spiral galaxies. If two dark matter particles annihilate in a dwarf galaxy, they can give rise to a large amount of energetic gamma rays, and the Fermi-LAT telescope can, in principle, capture these and provide a dark matter signal. So far, alas, no conclusive evidence for dark matter has emerged from these experiments.

As physicist Priscilla Cushman, involved in the Super CDMS dark matter search experiment, aptly described it:

> "The problem with all of this is, you're looking for something, a needle in a haystack. To build a detector, you need to make some assumptions about how shiny the needle is, how big it is, and where the haystack states you should start looking first. So that's what we do when we build these detectors. We think of the most likely place, and of course, we could be wrong. And so, as time moves on, and we don't find the needle where we expected to find it, we look further and deeper into the haystack."

There could also be other ways to look for a "break-up" of dark matter. A possibility is that in some parameter and mass range, dark matter could accumulate inside astrophysical bodies, such as the Sun or planets like Jupiter. Once they have accumulated in sufficient numbers, they could annihilate into photons or other secondary particles, such as electrons and positrons, which would be visible in detectors. For example, there was a Jupiter mission, the Galileo probe, which was launched in October 1989 by NASA and was capable of detecting charged particles in Jupiter's environment. Galileo was the first spacecraft to orbit an outer planet, and it orbited Jupiter for 8 years, discovering many important facts about Jupiter and its moons; however, it did not detect any signals attributable to dark matter.

4.9.4. *Searching for the Skinny Axion ("Sniff It")*

The axion discussed earlier in this chapter has strong theoretical motivation, since it solves a major problem associated with otherwise successful QCD theory of nuclear forces. It is fundamentally different from the dark matter candidates discussed so far: it is a wave-like dark matter due to its low mass. It cannot, therefore, be searched for using the "shake it," "break it," or "make it" techniques. Instead, the search strategy is what we refer to as "sniff it."

The axion has the distinctive property that its dominant interaction is with magnetic and electric forces (Fig. 4.6). A light wave of any frequency contains both oscillating electric and magnetic fields. If the axion dark matter in the universe is in an intense magnetic field, it can convert the magnetic field into a microwave electric field (an electric wave with a lower frequency than light), which can be observed in the laboratory [31].

There are several experiments that use this principle to search for axions. So far, no evidence for the axion has been found. Some of the leading experiments include ADMX (at the University of Washington), HAYSTAC (at Yale University), and CAPP (in Korea). The first search for dark matter axions using this technique was carried out by a collaboration involving Brookhaven National Laboratory, University of Rochester, Fermilab, and the University of Florida.

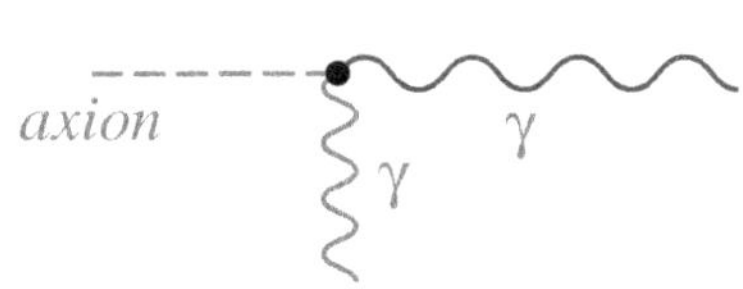

Fig. 4.6. The basic principle behind axion detection experiments. The axion (dashed line) can convert into two photons (wavy lines). These photons may equivalently be interpreted as oscillating electric and magnetic fields.

There are also other experimental proposals, such as the CASPEr experiment, which uses the nuclear magnetic resonance techniques (the same principles underlying MRIs in medicine), the CAST experiment, which searches for solar axions through their conversion into photons in an intense magnetic field, and the ABRACADABRA experiment at MIT, which employs sensitive detection of oscillating magnetic fields induced by axion dark matter.

More recently, there have been multiple proposals to use quantum sensing technologies to search for axion dark matter. The US funding agencies are increasingly investing in quantum sensing research, both for its potential in fundamental physics and its broader societal applications.

People have entertained the possibility that axions may not constitute the entire dark matter content of the universe but only a small fraction of it. In that case, one could still search for them. The aforementioned CAST experiment at CERN can look for axions produced in the Sun, regardless of whether they make up the dark matter of the universe. When X-rays interact with the electrons in the Sun, they can produce low-mass axions, and the CAST experiment searches for signals of these axions by decaying into light in a massive magnet on Earth that uses a magnetic field of about 100,000 times the Earth's magnetic field.

There is another kind of experiment conducted at DESY, the particle physics research laboratory in Germany, focusing on the search for axion-like particles (ALPS), where strong electric and magnetic fields are used to produce an axion on one side of an opaque wall. They are then allowed to pass through this wall. Since axions have very little interaction with matter in the wall, they will simply pass through it without obstruction and reemerge on the other side. There, they could decay to produce light. In this case, it would appear as though light is emerging from nowhere, since no light can pass through the opaque wall. Such setups are therefore called "light shining through the wall" experiments.

There are many other imaginative experimental approaches to searching for axions, most of which rely on their coupling to electric and magnetic fields to produce observable effects. Unfortunately, no convincing evidence for axions has been uncovered yet in any of these experiments.

Thus, the search for different kinds of dark matter continues and has become an increasingly active field, motivated by the fundamental question of the nature of the roughly 85% of the matter in the universe. For an excellent review of the current status of the searches and future directions, see Ref. [36].

Part II

Chapter 5

Why Mirror World?

"How often have I said to you that when you have eliminated the impossible, whatever remains, however improbable, must be the truth?"
— **Sherlock Holmes** in *The Sign of the Four* (Doubleday, p. 111)

The mystery surrounding dark matter is deepening. We know it is out there, but we do not know what it is. Despite a sharp uptick in efforts over the past two decades to discover it using multiple sophisticated techniques, all the conventional dark matter candidates described in the previous chapter, such as supersymmetric WIMPs or axions, seem to have escaped detection. This suggests that the story of dark matter may be more complex than currently understood and may yet hold many surprises. The mirror world or the replicate universe as a model for dark matter, described in the rest of the book, offers a novel approach to resolving this emerging puzzle of dark matter physics: that dark matter remains completely invisible to search methods used in our world and yet pulls stars and galaxies through gravity, thereby stabilizing the universe.

What is the mirror world, and why was it necessary to postulate it? What is its relevance to the problem of dark matter? As discussed in the following, the mirror world hypothesis suggests the existence of an identical duplicate of our world, with corresponding particles and forces. However, unlike the world we see around us, it is invisible to us. The most stable particles in the mirror world can therefore play the role of dark matter and escape detection. The only force common to both worlds is gravity, which explains why dark matter attracts matter in our world and influences the structure and stability of our universe.

For every particle in our world, there is a mirror partner, and all the forces in our world similarly have their mirror counterparts. The forces in the mirror world act only on the mirror particles and not on particles in our world, while the known forces in our world act only on particles familiar to us. *The only exception to this rule is the gravitational force, which is common to both worlds. Thus, all matter, including mirror matter, experiences and exerts gravitational force on each other.* This helps explain the effects of dark matter on our universe, since dark matter manifests itself through its gravitational influence on familiar matter in stars and galaxies. However, since the gravitational force is super-weak, experiments using detectors in our world to discover dark matter lack the sensitivity to detect its effects and have therefore failed.

The idea of a mirror world arose from the consideration of symmetries in particle physics. We discussed the examples of symmetries in Chapters 3 and 4. Let us recap some of that discussion as it applies to the mirror world.

5.1. Mirror Symmetry and Mirror Universe

As discussed in Chapter 3, symmetries are intimately intertwined with our understanding of the subatomic world. They have illuminated and shaped theories of atoms in many ways. The spectra of atoms were the first to reflect underlying symmetries, as did many other sub-atomic phenomena. A rule governing atomic spectra known as the Laporte-Meggers rule, discovered in the early days of atomic physics, states that light emission from atoms involving two energy levels with the same mirror reflection property is forbidden. This rule, published by Otto Laporte and William Meggers in 1925, was found to apply to all atoms.

More symmetries of different kinds were subsequently discovered in the subatomic world, ultimately leading to the birth of the standard model of forces and particles discussed in Chapter 3. Another important symmetry was Heisenberg's isospin symmetry, which was proposed to explain why the proton and the neutron had nearly the same mass, as discussed earlier. Extensions of isospin to higher symmetries led scientists to suspect that protons and neutrons were not elementary but contained "hard" constituents, called quarks. The quark picture of protons and neutrons was eventually confirmed by experiments. From there, physicists developed theories of the weak and strong forces and ultimately the successful standard model of elementary particles (Chapter 3). This spectacular improvement in our understanding of matter — from a vague and incomplete picture of particles to a well-defined and precise theory — was a monumental achievement of science in the second half of the 20th century. It raised the hope that an ultimate understanding of the laws of nature might be within reach, with symmetries playing a key role in achieving it.

A question that has always intrigued scientists is: How widespread is the influence of symmetries? Could symmetries be much deeper and more pervasive than those found in the subatomic world, extending to unfamiliar domains? We know that symmetries appear at the macroscopic level in animals, plants, and flowers. But could there be a symmetry at an even higher level, namely, the level of our universe? Such an idea might seem to defy common sense, yet this in fact was the ultimate consequence of a bold conjecture about our universe proposed in 1956 by two young scientists, Tsung-Dao Lee and Chen-Ning Yang.

Lee and Yang were trying to resolve a major puzzle in subatomic physics in the 1950s, and in the process, were led to postulate that there might exist a symmetric partner to our universe, invisible to us. Over the following years, this idea grew and received various refinements from physicists, eventually coming to be known as the mirror universe theory, which we describe in the remaining part of the book.

Clearly, the Lee–Yang conjecture was so revolutionary that if true, it would have far-reaching consequences for our understanding of the universe and possibly our existence and evolution. At first glance, it appears to defy common sense. When we talk of a symmetry, we normally examine the symmetric object in its entirety and discern the symmetry by studying it from different angles. In this case, however, we don't even observe more than a small part of our universe, let alone another one similar to it. Yet, according to this conjecture, there exists a symmetric counterpart to our universe. This is therefore a remarkably bold suggestion indeed, which is exactly why, for a long time, the idea remained only a footnote in the development of particle physics and cosmology. Only in the last three decades has serious research into this possibility gained momentum. In this book, we explain this idea in detail and explore its various implications. To appreciate why such a conjecture arose, we must first review a bit of history.

Let us examine mirror symmetry in a little more detail. It is a common form of symmetry found in many living and non-living objects. Remarkably, it also appears at the level of the atoms that make up living beings. Take, for example, the human body. If we draw a vertical straight line from head to toe, both sides of this line look identical, as if reflected in a mirror. In this sense, the human body is mirror-symmetric. To analyze this idea further, imagine a person wearing identical earrings in both ears. When she looks in the mirror, the reflected image appears essentially the same as the original. However, if she wears an earring only in her left ear, the mirror image will show the earring in the right ear, and the image is no longer identical to the original. The first situation is parity-conserving, whereas the second is parity-violating. It has been well known for a long time that processes involving nuclear, electromagnetic, and gravitational forces look the same as their mirror-reflected image, similar to the person with both earrings. This is why these forces were said to respect *parity symmetry*, another name for mirror symmetry. However, weak processes such as beta decay turned out to be different; they do not look the same when reflected in a mirror, as already noted in Chapter 3, hence they do not obey mirror symmetry. This surprising property of weak forces was not known until the mid-1950s.

What led scientists to suspect that weak forces are not mirror-symmetric? All elementary particles discovered were believed to possess a definite property called *intrinsic parity*. Using this property, physicists could test parity symmetry in most elementary particle processes in which the particles participated. For example, protons and neutrons have even intrinsic parity, while the pion has odd intrinsic parity.

Until the 1950s, all known elementary particle processes were found to respect parity symmetry. Then, accelerators began producing new particles, including K-mesons and π-mesons. Experiments involving the K-mesons revealed something puzzling. Their decay rates were relatively slow, indicating that they were believed to be caused by weak forces.

A detailed study of the K-meson decays soon revealed a problem. The K-meson and the π-meson are known to have odd intrinsic parity. However, the neutral K-meson was found to decay into two different sets of π-mesons: two as well as three π-mesons. A simple counting of parity showed that the parity of the two-pion state was even, whereas that of the three-pion state was odd. Therefore, one final state would be of opposite parity to the other, yet both appeared to originate from the decay of particles with the same mass and the same spin. If parity were a good symmetry as was widely believed, then only one of the decays should have been allowed, while the other should have been forbidden. Thus, there seemed to be a problem.

To avoid this problem, physicists initially assumed that there were two different particles, called τ and θ, which had the same mass and the same spin but opposite parity values, even and odd. The particle with even parity would decay into two pion states, while the one with odd parity would decay into three pion states. In this way, parity symmetry of the weak forces, which was responsible for these decays, would be obeyed. However, it seemed rather strange that two distinct particles should have exactly the same mass and the same spin. No such situation had been encountered before in physics. This problem became so acute that it acquired a name and came to be known as the τ–θ puzzle.

To solve this puzzle, Lee and Yang proposed in 1956 that difficulty might lie in our blind belief that all the force laws of nature obey parity symmetry. They suggested that the weak force, which is responsible for such important phenomena as the emission of light from the core of the Sun that sustains life on Earth, may not respect this symmetry. The same thing should happen with all weak force-mediated decays, for example, in the decay of the particle, the K-meson (then called τ and θ), parity symmetry would not be respected. If parity symmetry is not respected in the decay of the K-meson (or the τ and θ particles), a single

K-particle decaying into two different kinds of states would be allowed, and it would provide a simple resolution of the τ–θ puzzle. τ and θ would be one and the same particle, as common sense would dictate. They published this revolutionary idea in a 1956 paper in the journal *Physical Review.*

At that time, this idea was considered so radical that many physicists did not believe it. In fact, famous physicist Wolfgang Pauli is believed to have said the following upon hearing about parity violation [37]:

> "I refuse to believe that God is a weak left-hander......
>
> Physics is very muddled again at the moment; it is much too hard for me anyway, and I wish I were a movie comedian or something like that and had never heard anything about physics!"

Not only was Pauli skeptical, but Lee and Yang themselves were uneasy about their own suggestion. To challenge the deeply ingrained belief that parity was a good symmetry of all laws of nature must have bothered them immensely, so much so that they wrote down in the paper that parity could still be a good symmetry of nature if one modified the conventional interpretation of this symmetry. They suggested that there might exist two kinds of protons: left-handed protons, p_L (spinning opposite to the direction of their motion), and right-handed protons, p_R (spinning along the direction of their motion). Taken together, these two kinds could restore parity symmetry, even though only one of them was visible, while the other remained invisible. There was no explicit mention of a mirror universe in their paper in the way the concept is discussed today.

In fact, they even contemplated the possibility of one kind of proton, p_L (*L* standing for left-handed), oscillating into a right-handed proton, p_R. Such a possibility would be forbidden in the mirror universe picture as it is understood today. Nevertheless, their one-line mention of oscillating p_L to p_R contained the seed of the mirror universe idea. The first paper to present the modern picture of the mirror universe appeared 10 years later, in 1966, and was written by Lev Okun and collaborators [38].

In the modern version of this idea, parity symmetry would take us from one universe to another — the mirror universe (or the replicate universe, as we will call it here) — where the opposite parity states reside. In this picture, the missing side of the neutrino in weak decays would appear in the mirror sector. The response to Pauli's remark would then be that "God's weak right hand lies in the mirror sector," thereby restoring mirror symmetry of all interactions.

How does the mirror world idea restore parity symmetry? When we think of parity symmetry, we literally imagine the reflected image of an object and see its presence in the same space as the left side of a human going together with the right side. According to the new picture, however, the reflected part of one side was not located in this visible world but in a fictitious world called the mirror world. When both worlds are considered together, "parity" symmetry is preserved. However, when viewed from within a single world, parity appears to be a broken symmetry. For example, this happens in the weak processes in our world when a neutron decays into a proton, an electron, and a neutrino. As discussed earlier, in this neutron decay, only one side of the neutrino appears. According to the new mirror universe idea, the other side of the neutrino would appear in a separate neutron (the mirror neutron) decaying in the mirror world. In this way, parity symmetry would be restored when both worlds are considered together (Fig. 5.1).

As shown in Fig. 5.1, when we view things within the framework of two worlds together, "parity" symmetry is preserved. However, when

Fig. 5.1. Illustration of parity symmetry and the mirror world idea. Left: A triangle with full visible symmetry in our world. Right: Parity transforms one half of the triangle in our world into the corresponding half in a mirror world. Since the latter resides in a different world, we cannot see it, and this automatically brings in two worlds (or universes) into the picture.

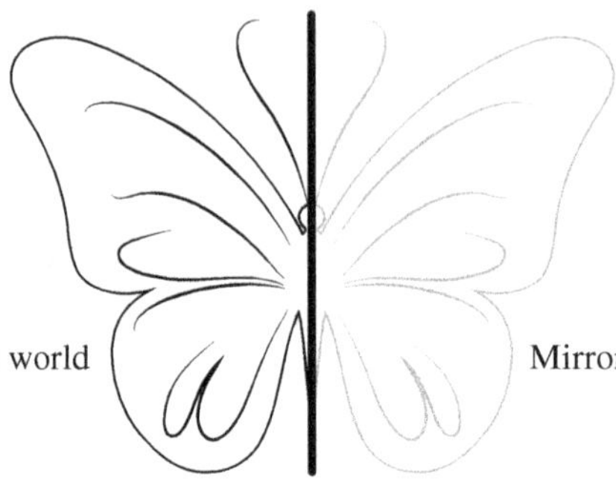

Fig. 5.2. Illustration of the mirror-world idea using a butterfly analogy. A butterfly with only a left wing in our world would appear mirror asymmetric. However, if the other half of the butterfly with the right wing is in the mirror world, then both worlds taken together will restore reflection symmetry of the whole butterfly. God would then not be a weak left-hander but would possess mirror-symmetric hands, with the left and right hands residing in different universes.

we only observe things in our world (or the mirror world) alone, weak forces would not be parity symmetric. The situation is analogous to a butterfly whose left half lies in this world, while the right half resides in an invisible world. We call this invisible world the mirror world or the replicate universe (see Fig. 5.2).

The obvious question that arises then is: What happens to the forces such as strong and electromagnetic ones, which are known to be mirror symmetric under mirror reflection within our world alone? According to this new theory, there would simply be two different kinds of mirror-symmetric strong forces and electromagnetic forces: one operating in our universe and another in the mirror universe. Each would be separately symmetric under the usual mirror reflection within its own universe. This suggests that there are two kinds of mirror reflections: the first is the familiar kind, in which the mirror reflected image resides in our world, and the second kind is the one in which the mirror reflected image of a process in our world resides in the other world. Under this second kind of mirror reflection, which can be called "super-mirror," all particles in our world transform into particles in the second world, and vice versa. The reflected particle is called the mirror partner of the particle in our world. We do not observe any of the super-mirror-reflected particles (or mirror partners) in our atoms or accelerators. Clearly, therefore, to reconcile this with common sense, one is naturally led to the idea that two universes coexist. The forces in the mirror universe — except gravity — are fully invisible to us, just as those in our universe will be fully invisible to the hypothetical "inhabitants" of the mirror universe. Thus emerges a two-universe model from aesthetic considerations together with the observed scientific property that weak forces do not respect mirror symmetry. This two-universe framework forms the central theme of this book.

It is important to note, however, that the mirror universe is not the only framework proposed to understand the breaking of mirror symmetry by weak forces. Nearly 20 years after the Lee–Yang proposal of parity violation and possibility of the mirror universe, in 1974–1975, another more economical approach to reconcile parity invariance with observed parity-violating weak forces was proposed by R. N. Mohapatra (the author), Jogesh Pati and Goran Senjanovic [23]. These are the left–right symmetric theories, already mentioned in Chapter 3. In these theories, the particle mediating the mirror counterpart of the observed weak force (the W_R boson) was assumed to be much heavier. As a result, the interaction strength of the parity-reflected weak interactions is suppressed at lower energies, keeping the right-handed interactions invisible. However, when machines with energies higher than the heavy mass of the W_R are available, the right-handed weak force can become visible. This is to be contrasted with the mirror world scenario, where the mirror counterpart of the observed weak interaction is hidden in the mirror world and can never be directly probed by colliders.

The connection between the heavy mass related to the unobservability of the mirror weak interactions in the W_R scenario and the lightness of the neutrino was demonstrated by Goran Senjanovic and the author in 1979 [39]. The fact that neutrinos are extremely light has now been firmly established through many painstaking experiments over the past three decades. The suppression of these new parity–counterpart interactions is achieved in a mathematically consistent way through the mechanism of spontaneous symmetry breaking, which has been used so successfully in constructing the standard model. In this approach, no new universe needs to be invoked. This theory predicts a new particle, W_R, the right-handed partner of the well-studied W-boson that mediates the familiar weak interaction. Thus, a heavier counterpart of the neutrino and a heavier counterpart of the familiar W-boson form the two key cornerstones of the non-mirror world way of understanding mirror symmetry (or parity) breaking in weak forces. The Large Hadron Collider as well as many other experiments are looking to discover these two particles: the W_R and the right-handed neutrino. This search will continue when planned higher-energy colliders are built in Europe and/or China.

In the left–right symmetric model, the neutrino has the additional property that it can be its own anti-particle. This allows other low-energy experiments that do not involve an accelerator to search for evidence in favor of this scenario. It looks for neutrinoless double beta decay of nuclei, e.g., the Ge^{76} nucleus decaying to the Selenium (Se^{76}) nucleus plus two electrons or the Xenon (Xe^{136}) nucleus decaying to the Barium (Ba^{136}) nucleus plus two electrons. There are no neutrinos in the final state of these processes, hence the name "neutrinoless double beta decay." The fact that neutrinos are their own anti-particles makes the two neutrinos emitted in the double beta decay "eat" each other. There are a whole host of other nuclei that can also undergo neutrinoless double beta decay. These decays are sensitive to the existence of W_R and the heavier partner of the neutrino in left–right symmetric theories. These experiments are being carried out to establish that neutrinos are their own anti-particles as well as to search for the effects of the right-handed W_R and heavier right-handed neutrinos. These are very rare processes expected to occur once every 10^{27} years or so. No evidence has so far been uncovered for this process, but this remains a very active field of research, with many ongoing experimental efforts.

In this book, we focus on the first approach — the "mirror universe" way of understanding parity asymmetry in weak forces — and the associated new cosmological and astrophysical ideas that emerge from it.

5.2. From the Neutrino World to the Mirror World

This idea of a mirror universe was developed into a full theory over the half-century, following the suggestion in the Lee–Yang paper, by many physicists [40–42, 44]. There has been a surge of activity in this field during the past two decades, especially in connection with dark matter. We will see what it has to do with dark matter in the following.

The basic picture is that the standard model and a mirror partner of the standard model coexist in the same space–time. For every quark and lepton in our world, there is a mirror quark and mirror lepton in the mirror world. For every kind of exchange particle that leads to various forces in our world, there are mirror partners of these force particles whose exchange, only between the mirror particles, generates mirror forces among them, just as in our sector of the universe. There is therefore a complete one-to-one correspondence between our forces and particles and mirror forces and mirror particles. The strengths of the forces are the same if the masses of the exchange particles are the same. This is just like what we see in a mirror, where the reflected image looks identical to the original.

The only thing to remember is that, whereas both left and right sides of all quarks and the electron exist in each universe, the mirror neutrino has only one side in our universe, with its other side residing in the mirror sector. This, after all, was the starting point of the mirror universe idea. Obviously, the particles of our world (also sometimes called the visible world or the visible sector) have all been experimentally observed, whereas none of the mirror sector ones have been, except possibly the dark matter particle, as we discuss in the following.

The mirror quarks and leptons exchange only the mirror force particles $\boldsymbol{W', Z', G', \gamma'}$ to generate mirror forces, whereas our quarks and leptons exchange only our $\boldsymbol{W, Z, G, \gamma}$ particles to generate the known forces. The prime notation distinguishes the mirror sector exchange force particles from those in our sector.

While mirror symmetry dictates that the force strengths in the two sectors are originally the same at the beginning of the universe, they could change during cosmic evolution, as noted earlier. As a result, they may now be very different or slightly different, depending on the nature of the universe's evolution and how it affects mirror forces. We discuss this later.

Do the two sectors communicate at all? At the basic level, there is no cross-talk between these two sectors except a very weak one via the gravitational force, since gravity acts on all matter and is common to both sectors. In other words, mirror particles, in addition to exerting gravity forces among themselves, also exert gravitational forces on our particles, and vice versa. Consequently, the presence of mirror particles will affect the motion of stars and planets in our universe, and vice versa, but otherwise produce no observable effects, i.e., create no other disruptions, so to speak. It is then clear that a stable particle from the mirror world can behave as dark matter in our universe, since it satisfies all other properties that dark matter needs to satisfy: it influences the motion of stars and galaxies but does nothing else. One important implication of this possibility is that, without adding any extra forces to the theory, there will be no way to detect this kind of dark matter in ongoing laboratory or collider experiments. This is again because our detectors are sensitive only to the forces present in our universe, whereas dark matter is sensitive only to the mirror forces.

It may, however, be possible that the particles in the two worlds communicate through ultra-weak forces other than gravity. In subsequent chapters, we discuss what implications such variations of the original picture could have for astronomy as well as for its possible laboratory detection of mirror partners.

Gauge bosons of each sector are produced only by particles of the same sector. This is why we cannot see what is going on in the mirror world, and neither can the mirror particles "see" what is going on in our world. In our "world," we can observe objects only using our kind of photon, and similarly, in the mirror world, objects can be observed only using the mirror photon. The mirror photon, incidentally, could be the dark photon discussed in Chapter 4, whereas our photon is the light we use in our everyday life. The electricity flowing in our homes is different from the electricity flowing in "the homes" in the mirror sector.

While at the basic level, there is complete symmetry, there are two aspects to what we observe in the universe today. First, there are the constituents and forces that form the basic underlying framework, as discussed in Chapter 3. The second aspect is the collective evolution of these constituents and forces from the early moments of the universe till now, when we make our observations. During this process of cosmic evolution, particle properties can change, and similarly, forces can also change both in their strength and in their range (i.e., how far the influence of the force extends), thereby influencing what goes on around them.

Ideally, if both our universe and the mirror universe evolved in a completely identical manner, what we see in our universe today would be a complete duplicate of the mirror universe with only minor differences. There would be a mirror Sun, a mirror Earth, mirror humans, mirror plants, and even presumably mirror winds blowing past mirror trees when winds in our backyard sway a tree. They will be exactly in the same locations as in our universe. However, as we will argue later, this is a very unlikely situation, since the basic laws of forces being the same does not guarantee that the evolution of the universe will remain identical. Thus, what we observe in our universe is expected to be very different from what is going on in the mirror universe, where evolutionary effects may lead to significant differences. All we can conclude is that, since the weak forces appear parity asymmetric only in experiments today, the mirror universe hides the parity partner process involving weak forces that we observe.

Strictly speaking, however, we do not know exactly what happened as the two sectors of the universe evolved. Many things could be changing in the environment of the two sectors governed by things that do not affect what we see today. Thus, the best we can assume is a complete mirror symmetry in basic constituents and forces at the very beginning of the universe.

Depending on how much the evolution of the universe affects the masses and interactions of different particles, we can envision two possible mirror universe scenarios: a symmetric scenario, where the strengths of all forces and masses of particles remain the same in both sectors, and an asymmetric scenario, where the strengths of forces as well as the particle masses become different in the two worlds. In subsequent sections, the first one will be referred to as the symmetric mirror world scenario and the second as the asymmetric mirror world scenario. Their observable consequences are very different, and similarly, ways to test for them will also be very different. Some consequences, though, can remain the same. The symmetric mirror world scenario has been actively pursued by Robert Foot and Ray Volkas of the University of Melbourne and Zurab Berezhiani and his collaborators at the Gran Sasso Institute in Italy. The author has found the asymmetric mirror models to be particularly interesting and has collaborated on this topic with Zurab Berezhiani, Shmuel Nussinov, and Vigdor Teplitz.

It is important to emphasize that the replicate universe idea is different from the many-universe idea in quantum mechanics or the multiverse idea in string theory. In the latter cases, even space–time may be different in different universes, and the laws of physics in different universes are not necessarily related.

In this part, we focus solely on the two-universe idea, where the laws of physics are the same in both universes in the early moments of cosmic history but become different later.

One can visualize the two mirror worlds as two families living on two opposite sides of a wall; they cannot see each other, and as a result, they just keep guessing what the other family would be like. The more information one family has about the habits and activities of the other, the more accurate their guesses would be. This is the situation for our mirror world predictions. We can only make educated guesses as much as possible until concrete, direct evidence for mirror sector physics is discovered.

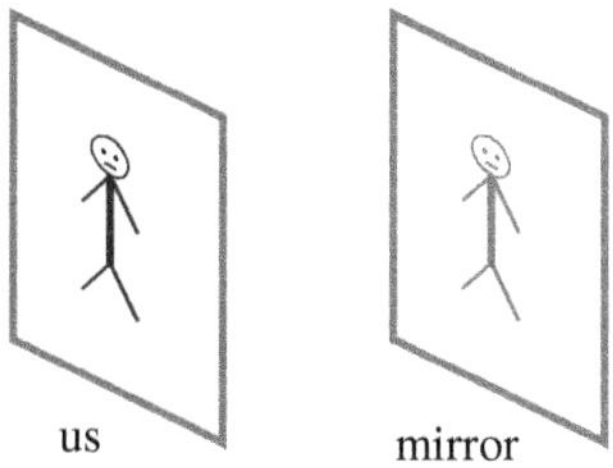

Fig. 5.3. M-theory picture of the mirror world. Each "area" represents a 1+9-dimensional brane world.

Finally, "parity invariance" is not the only motivation for entertaining the mirror universe idea. String theories also often predict this kind of a symmetric two-universe picture for elementary particles. In the so-called M-theory version of string theory, there exist space–time areas called D-branes, which are multi-dimensional spaces. Each of these branes can behave like a separate universe with its own 10-dimensional space–time, as discussed in Chapter 1. There are usually two 10-dimensional D-branes with a symmetry that takes one brane to the other. M-theory is an 11-dimensional theory (one time plus 10 space dimensions), and each of the branes (or universes) starts out as a 10-dimensional universe (one time + nine space). Once the extra dimensions curl up, the mirror world picture we have here can emerge. The standard model and its exact counterpart, the mirror standard model, can then reside on the two branes exactly like in the mirror universe picture (Fig. 5.3).

5.3. Symmetric and Asymmetric Mirror Worlds

New ideas often open the door to new possibilities, and the mirror world idea is no exception. Having outlined the general reason for considering a mirror universe scenario as a serious possibility, we can now discuss the different kinds of new phenomena that may arise within this framework. The various implications suggest a variety of observational tests. If evidence for a mirror universe is eventually discovered, such observations could gradually reveal the detailed nature of that universe.

From a theoretical point of view, as emphasized, there are two possible broad classes of mirror world scenarios: the symmetric mirror world and the asymmetric mirror world. Before discussing the detailed implications of each possibility, let us first examine how the two scenarios can arise. To understand this, let us recall the discussion of the standard model in Chapter 3. The most relevant point in our discussion here relates to how the particles acquire mass in the standard model and the role played by the Higgs boson. As explained earlier, the masses of the standard model particles — quarks, leptons, and gauge bosons — arise through the Higgs mechanism. In this mechanism, the vacuum state (or the lowest energy state of this model) contains a constant value of the Higgs field spread throughout the universe. This is referred to as the Higgs spread in this book. The Higgs spread plays a crucial role in the standard model. Since all the particles of the standard model, except the neutrino, interact with the Higgs boson H, they experience the presence of this background field. An analogy that may help illustrate the idea is that of a fish that swims on the surface of a pond. It not only feels the ripples on the surface of the pond but also experiences the still surface itself. In a similar way, a particle that scatters off this ground state (or the Higgs spread) as it propagates will feel the spread, and its effect gives the particle its mass. Any particle like the photon (light) or the neutrino which do not interact with the Higgs spread will remain massless. The actual mass of a particle that interacts with the Higgs field will depend on the strength of the force with which the Higgs field connects to the spread and the value of the Higgs spread. Since the spread has the same value for all particles in the standard model, the different masses of standard model particles arise from the different strengths of the Higgs coupling (Fig. 5.4).

In the mirror model, there are two Higgs bosons: one associated with our world (H) and another associated with the mirror world (H').

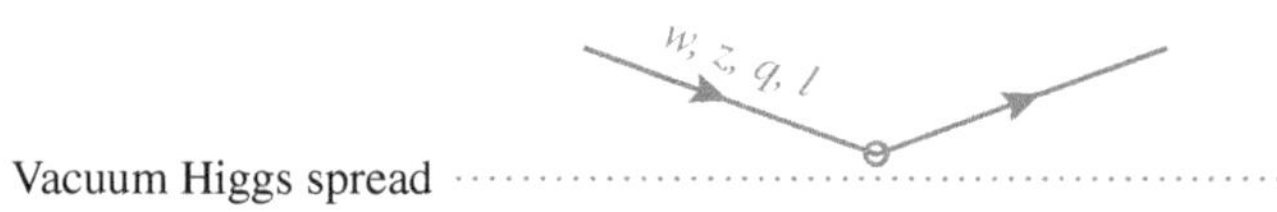

Fig. 5.4. Illustration of how particles acquire mass in the standard model through the Higgs spread. The stronger the force, the larger the mass of the particle.

The coupling strengths to both the Higgs fields will be identical in both sectors due to parity symmetry. However, the vacuum spread in both sectors need not respect parity symmetry and can be either the same or different. If the spreads are identical, the particles of the mirror world will have exactly the same masses as their partners in our world. This model is called the symmetric mirror world model. In contrast, if the vacuum Higgs spread is different in the two sectors, the particle masses will also be different, and this model is called the asymmetric mirror world model.

5.4. Mirror Matter as Dark Matter

From the above discussion, it is clear that, just as in our world, mirror quarks in the mirror world will combine with one another at a certain stage in the evolution of the universe to form mirror protons and neutrons (denoted p' and n', respectively). As in our universe, the mirror proton p' could be lighter than the mirror neutron n'. However, since forces could change during the evolution of the universe (even if they started out being identical at the beginning), it is also possible that the mirror neutron n' is lighter than the mirror proton p'. The mirror proton would combine with the mirror electron to form the mirror hydrogen atom (denoted by H'). In either case, whichever particle is lighter — H' or n' — would be stable. If produced in sufficient abundance, these particles could constitute the dark matter of the universe.

The produced dark matter will then seed the formation of galaxies and stars in both sectors of the universe. This will occur when the gravitational force of a clump of matter exceeds the outward pressure of the moving matter inside it as the universe cools down. Considerable theoretical effort is currently devoted to studying the possible observability of dark stars, dark galaxies, and dark black holes predicted by the mirror matter models [43–46].

Chapter 6

Cosmic Evolution and Mirror Universe

> "The Big Bang theory says nothing about what banged, why it banged, or what happened before it banged."
>
> — **Alan Guth**

A matter of deep curiosity to all living beings is the nature of the environment they inhabit. For humankind, the immediate environment consists of the Earth and the Sun, which together are part of the Milky Way galaxy, which is but a small part of our vast universe. This curiosity has driven humans for centuries to gather knowledge not only about the solar system and our galaxy but also about the universe as a whole.

After decades of research, it is now almost conclusively established that our universe began as a nearly point-like dense fireball. It initially underwent a phase of very rapid expansion, called the inflationary expansion, followed by a slower phase later, called the Hubble expansion. This expansion continues till today, with an important modification: It began accelerating "recently," i.e., about 5 billion years ago. This forms the broad framework of the Big Bang model. Two key observational pillars provide confirmation of this scenario: (1) the discovery of the cosmic microwave background (CMB) by Arno Penzias and Robert Wilson in 1964 and (2) the confirmation of Hubble's law, which states that the speed with which the galaxies recede from us is proportional to their distance. Once we entertain the possibility of a second, or a mirror, universe, new questions naturally arise: How would this expansion picture be affected? How would the mirror sector evolve and reach its present state? At present, the existence of such a mirror universe of course remains to be confirmed. Nevertheless, one can still outline the salient features required for such a scenario to be consistent, independent of the details.

An important step in understanding a mirror universe is to first examine how our universe evolved from the earliest moments after the Big Bang to the present day. Since a mirror universe if it is there, coexists with our universe, it is useful to first consider the evolution of our universe in isolation, i.e., without being accompanied by the mirror sector. We can then explore what would have happened if the mirror universe coexists with it.

Our universe is nearly 14 billion years old, and we possess substantial knowledge of its history from about 1 minute after its birth to the present. What occurred before that remains largely speculative, although some of these ideas are very plausible and may turn out to be correct. Let us recap what we know so far.

6.1. The Standard Big Bang Theory of Our Universe

As the twentieth century was unfolding, and as physicists were expanding their knowledge about the subatomic world, cosmologists were trying to improve their understanding of the universe and were developing a theory of its structure and evolution.

Belgian priest and cosmologist Georges Lemaître was the first to propose the Big Bang theory in 1927. According to this, the universe had started off as an extremely hot fluid in a minuscule volume and had been expanding and cooling ever since. This agrees with the Hubble expansion and was also predicted by Einstein's theory of general relativity, as shown by Soviet physicist Alexander Friedmann in 1922 and later by US physicist Howard Robertson and UK physicist Arthur Walker in 1935 and 1937, respectively.

An immediate implication of this idea is that the universe had a beginning, which is contrary to the Aristotelian belief that the universe had an infinite past. The Aristotelian view could be tied to the so-called steady-state theory of the universe, according to which the universe is continuously expanding without a beginning or an end while maintaining a constant average density due to continuous creation of matter. This theory was pioneered by British astronomer Fred Hoyle, Herman Bondi, and Tommy Gold in 1948.

By the mid-1950s, two competing theories of the universe were therefore going head-to-head: (i) the steady-state theory and (ii) the Big Bang model. As more observations were made, three pieces of evidence strongly pointed toward the correctness of the Big Bang theory and ran counter to the steady-state theory: (i) the Hubble expansion, confirmed as the distances and speeds of more galaxies were determined; (ii) the calculation of the elemental abundance of hydrogen and helium in the universe when it was about 1 minute old; (iii) the 1964 discovery of the CMB by Arno Penzias and Robert Wilson. Other details began to be explored, and just as we now have an almost complete understanding of the observed matter and forces in the universe, our understanding of how the universe evolved from its beginning to its present state has also improved considerably during the last half-century, thanks to many cosmological observations and theoretical developments. Many details of the Big Bang theory have been confirmed. These are summarized in the following [47, 48]:

Astronomers have now established the age of the universe to be about 14 billion years. The Big Bang theory, which states that the universe was very hot and very tiny in the beginning, has been cooling and expanding since then and will continue to expand in the future. As it expands, it creates more space. At some point during this cooling period, galaxies, planets, and stars formed. Dark matter is believed to have played a major role in galaxy formation, as already noted in Chapter 4 and as is further discussed here.

All particles being hot, in the beginning, were moving extremely rapidly at or close to the speed of light. They packed a lot of energy, and as the universe expanded, the density of energy and particles decreased steadily. According to Einstein's theory of general relativity, the expansion rate of the universe at a given moment depends on the energy density. So, the lowering of the energy density led to a slowdown of the expansion rate. This is reflected in the Hubble law, which states that the speed of a particle (star, galaxy) is proportional to its distance from the observer. For instance, a star that is farther from the Earth moves faster than a star that is closer.

The slowing down of the expansion rate of the universe seems to have changed in the recent past, as a new accelerating phase was discovered in 1998. The latter came from a study of objects called type I supernovae. From the brightness of the type I supernovae, one can find out how fast they are moving. We still do not understand why this acceleration occurred, although there are some possible hints that a constant background energy, called the cosmological constant, first introduced by Einstein, could be the reason. Einstein called it his biggest mistake, but it turns out that it may indeed be there in the universe. If so, the value of the cosmological constant is very tiny, and how it got to be so tiny has puzzled physicists since the acceleration phase was discovered. Some people think it could be a near-massless particle called quintessence floating around in the universe, which may play a role in causing the acceleration. We still do not know the final answer, although recent observations by the ground-based Mayall Telescope at Kitt Peak National Observatory in Arizona, as part of the Dark Energy Spectroscopic Instrument (DESI), seem to be favoring the quintessence hypothesis. According to this, it would appear that the accelerated expansion could be slowing.

Coming to the evolution of matter, it is likely that all particles in the beginning stage of the universe were massless, although one cannot be completely sure. At a certain point, most of the particles, initially massless, acquired their masses due to a particle physics phenomenon called the Higgs mechanism, as discussed in Chapter 3. Once the temperature of the universe dropped below the mass of a particle, it began to move more slowly. If unstable, the particles decay into lighter particles at lower temperatures and disappear. If a particle has energy exceeding its mass and is unstable, then it will decay, but its decay products can recombine in the reverse process to recreate the particle. This state is referred to as *thermal equilibrium*.

During the expansion of the universe, several key events occurred that shaped its present structure. One of the major ones is the inflationary expansion that occurred shortly after the Big Bang. Other important milestone events in the history of the universe include the following: (i) creation of matter–anti-matter asymmetry; (ii) formation of protons and neutrons from quarks; (iii) Big Bang nucleosynthesis (BBN) which converted protons and neutrons to low mass nuclei, such as deuterium, helium, lithium, and beryllium, the basic building blocks of atoms; (iv) domination of matter over radiation around the time when the universe was about 50,000 years old; (v) formation of atoms from free electrons and nuclei, such as free protons, deuterium, and helium nuclei, which occurred when the universe was about 400,000 years old; (vi) growth of small and large structures, seeded by dark matter, into galaxies and stars, beginning when the universe was about 700 million years old; (vii) slow accelerated expansion at a later phase, driven by the dominance of the cosmological constant, beginning when the universe was about 9 billion years old. This expansion continues, drifting all galaxies apart from each other in a slow thinning down of the density of matter.

We must caution that observational evidence exists only for epochs later than about a minute after the Big Bang, i.e., from the time of BBN onward. Evidence for the earlier inflationary phase comes indirectly from precision measurements of the cosmic background radiation by observational devices, such as the Wilkinson Microwave Anisotropy Probe (WMAP) spacecraft, launched by NASA, and the Planck satellite, launched by the European Space Agency (ESA). Events between the end of inflation and the onset of nucleosynthesis time are mostly educated guesswork, quite likely to be correct but not as clear-cut. There could be alternative evolution scenarios capable of explaining the same observations.

6.2. Inflation

What is the inflationary expansion of the universe, and why is it needed? There are several reasons for its introduction. If the expansion of the universe was given by the Hubble rate all the way from the beginning of the universe till now, we would encounter several problems. The first is the observation that the CMB temperature is almost the same in every direction. This implies that all parts of the universe must have been in contact for a long time since the beginning, so that the temperature became uniform in all parts and directions. This is hard to understand in the Hubble expansion picture. Given information about the present size of the universe, we can extrapolate back to earlier epochs using the Hubble rate of expansion to estimate its size at earlier times. If the Hubble expansion were the only type of expansion, the universe at earlier times (such as during BBN) would have had a much smaller size than currently believed. There would then have been many disconnected regions that would later have to be "patched together" to form the universe at that time. This makes it difficult to understand how the entire universe could have a uniform temperature since there would have been no "overlap" between these disconnected regions. There were also other problems, such as too many heavy monopoles in the universe packed into a small space.

To resolve these problems, Alan Guth suggested in 1980 [49, 50] that the universe underwent a period of extremely rapid expansion in its early stages. This expansion dramatically increased its size, thus explaining the observed isotropy of the CMB. One small patch becomes the entire universe. This idea was further developed by Andre Linde, Andreas Albrecht, and Paul Steinhardt [51], who proposed a newer and more refined version of the model, explicitly discussing how inflation would end.

The following example provides a simple analogy to a causal connection through inflation. Imagine living in a small cubicle enclosed by walls within a large hall filled with similar cubicles. In such a setting, one cannot know what is happening in the distant parts of the large hall. However, if the walls are removed, one can see everything, know everything, and react to everything. Inflation is analogous to removing these barriers, effectively turning the entire hall (the universe) into one large open area.

Another major puzzle of the Big Bang theory is that the early universe was extremely smooth and uniform, whereas the present one is highly non-smooth and non-uniform, containing stars, galaxies, and clusters separated by vast, empty regions. This is like cheese made out of milk: Milk is uniform, but cheese is not. How did the non-uniformities emerge? The inflation theory also seems to provide a solution to this puzzle. During the period of inflation, some small local changes in density naturally developed, as would be the case in any fluid being constantly disturbed [52]. These fluctuations eventually grew to become galaxies and clusters observed today. This theory seems to match the overall features of the galaxy patterns in the universe.

After inflation ends, familiar and new kinds of matter are produced in a process called reheating, and the universe resumes Hubble expansion. Thus, the universe undergoes an initial inflationary phase followed by the standard expansion phase, which continues to the present.

The more recent accelerated expansion phase of the universe, mentioned earlier, is believed to have begun when the universe was about 9 billion years old. This late-time acceleration is distinct from the early universe inflationary expansion.

Despite its successes, the underlying mechanism behind inflation remains a mystery. Broadly speaking, it is believed to have been caused by an object similar to the Higgs boson of the standard model (or maybe even the Higgs boson itself). There are also speculations that it could be related to the existence of extra space dimensions. The true origin of inflation is still unknown. Could it even be that inflation is not the correct theory for the beginning of our universe? That is unlikely but possible. At the moment, however, inflation may be the best theory we have for understanding the universe prior to standard Hubble expansion.

6.3. Matter-Anti-Matter Asymmetry

The next important milestone in the history of the universe is the time when an asymmetry between matter and anti-matter was created. If we look at the universe now, we find that it consists of only matter (such as protons, electrons, neutrons, and nuclei) and no anti-matter (such as anti-protons, positrons, and anti-nuclei). Cosmic rays have been found to contain a small amount of anti-matter, which can be explained by their production from energetic proton–proton collisions in space. Extensive searches have established that there is no known place in the universe that contains anti-matter. In fact, if there were anti-matter in comparable abundance to matter in close proximity, there would be annihilation and explosion, making the universe completely empty of matter and full of radiation. Our existence proves that this is not the case. If there were anti-matter isolated in one area of the universe and matter in another, with an impenetrable wall between them, then there would be an asymmetry (anisotropy) in the universe that should be reflected in the CMB observations. However, no such asymmetry has been found after very careful measurement by different experiments. The amount of matter relative to radiation is very small indeed — about one part in a billion is matter, with essentially no anti-matter present.

So, if matter and anti-matter are equally plausible, where did all the anti-matter go? Could one argue that the matter was simply put in place by some unknown mechanism at the beginning, and that is what we observe today? This raises the question of who or what put the matter there? In any case, even if such a mechanism existed, that idea would be untenable if the inflationary picture is correct. Since inflation causes a huge, sudden expansion of the universe, whatever matter existed before would be diluted to a minuscule amount and cannot explain the matter we observe today. Inflation essentially "vacuum cleans" the universe of all pre-existing matter due to its rapid expansion. So, where did matter really come from, and at what point in the evolution of the universe was it created?

There are many theories to explain the matter–anti-matter asymmetry, and they generally correspond to different eras in the evolution of the universe and go by different names. The two most widely discussed ones are related to the mass of the neutrino and the origin of quark masses. Understanding even one of these in a simplified way provides insight into the origin of matter. This is an active field of research, and the final word on this is far from settled, but plausible scenarios have been proposed.

6.4. Phases of the Universe

What do we mean by phases of the universe? The basic idea of a phase can be understood through a simple analogy. When we heat water above 100°C, it boils and changes from liquid to steam (gas), and when we cool it below 0°C, it goes from liquid to solid ice. Steam and ice represent different phases of water, and the temperature at which this change of phase takes place is called the phase transition temperature. This is a common phenomenon for many liquids and solids around us and is a very well-understood subject in physics. When water becomes ice, tiny crystals of ice form, and when water becomes steam, the bond between water molecules breaks, allowing each molecule of water to fly away on its own. Thus, there appears to be a markedly different behavior of water molecules at the phase transition temperature. In a similar manner, as the universe cools from hot to cold, at certain temperatures, new phenomena occur, and drastic changes take place in particles in the universe. For example, around the time when the universe was 10^{-11} seconds old, the *W*-boson that mediates weak forces and the quarks of the standard model that constitute matter acquire mass; prior to this epoch, they were massless. A massless *W*-boson implies that weak forces can affect quarks and leptons separated by arbitrary distances, unlike in the present universe, where the weak force is only effective inside the nucleus, as in beta decay. The weak forces prior to this epoch behaved as long-range interactions, similar to electric and gravitational forces. The moment when this change of phase took place is called the electroweak phase transition point, much like the freezing point (water–ice transition) or the boiling point (steam–water transition).

Similarly, another phase transition took place when the universe was about a 10th of a millisecond old, when quarks combined to form protons and neutrons. Recall that at this point, three quarks combined to form protons and neutrons; a quark and its anti-particle combined to form mesons. Before this moment, quarks and anti-quarks were freely floating. This is called the QCD phase transition, since it refers to the QCD force, which was much weaker in the very early universe, becoming strong enough later to bind the quarks together.

The weak phase transition and the QCD phase transition described above are the phase transitions we are reasonably certain about. However, we do not know if there were other phase transitions at earlier times, since the detailed nature of theories beyond the standard model remains unknown. For example, in some beyond-the-standard-model theories, there can be hugely massive objects that can form at the phase transition time called "monopoles." If they are produced, they are stable and will hang around till now. However, we have found no evidence for monopoles so far from the many experiments carried out. In fact, if cosmic inflation occurred after such a phase transition, in which monopoles were produced, these monopoles would be effectively "vacuumed" away, which may explain why we do not observe them today.

6.5. Big Bang Nucleosynthesis and How Atoms Formed

Before the universe was 1 minute old, the number of neutrons and protons remained the same due to back-and-forth reactions involving the weak force that change a proton to a neutron, and vice versa. Due to the weakness of the weak force, somewhere near the time when the universe is 1 minute old, the Hubble expansion rate overtakes the rate at which weak forces convert neutrons into protons. After that time, weak interactions cannot change protons to neutrons as efficiently, and vice versa, since they become too slow compared with the expansion rate of the universe. The numbers of neutrons and protons at that point remain frozen. As the universe cools, these neutrons and protons undergo nuclear attraction and combine two protons and two neutrons to form helium nuclei. Calculations show that there are many more protons than neutrons at the time when the neutron and proton numbers freeze. Since it takes an equal number of protons and neutrons to make a helium nucleus, some protons are left over, and the rest become helium nuclei. The leftover protons later form hydrogen atoms, and the helium nuclei form helium atoms. This process is called Big Bang nucleosynthesis (BBN), and it leads to roughly 75% of the mass of the matter remaining as hydrogen and 25% becoming helium nuclei, with a small fraction of other heavier nuclei formed through further nuclear reactions. These include smaller amounts of lithium and beryllium, and also some small amount of deuterium, whose nucleus has one neutron and one proton. These are essentially all the nuclei formed during BBN, and their abundances persist until later stages of cosmic evolution. The basic theory of BBN was outlined by Ralph Alpher, Hans Bethe, and George Gamow in 1948 in the famous Alpha–Beta–Gamma paper.

BBN is one of the major triumphs of particle physics, nuclear physics, and cosmology. It predicts the abundance of deuterium and helium with remarkable accuracy and is in excellent agreement with observations (deuterium and helium are two of the lightest elements in nature). This is especially notable since the abundance of deuterium is about 10,000 times smaller than that of helium, yet the current BBN theory predicts it very precisely. The BBN model also predicts the abundance of lithium-7; however, this last prediction disagrees with observations by a factor of four. This discrepancy is considered one of the major puzzles that may point to new physics beyond the standard model (at least some people think so). The neutrino plays a key role in the formation of hydrogen and helium, the first elements in the periodic table.

It is worth emphasizing here that there is something magical about BBN in the following sense. The helium-to-hydrogen ratio depends on the Hubble expansion rate becoming faster than the weak interaction rate at the BBN epoch. The Hubble expansion rate depends on the number of relativistic species, including three neutrino species, whereas the weak interaction rate is a completely independent parameter that has nothing to do with the three neutrinos. The striking fact is that the number of neutrinos and the weak force rate are such that the correct amount of helium is produced. Why this coincidence occurs remains an open question in cosmology.

What happens after BBN is completed? At the time of BBN, no atoms could form since the temperature of the universe was about 10,000 million degrees, and at such high temperatures, all atoms break up to nuclei and electrons. The electrons are free and much too energetic to bind with nuclei to form atoms. The atoms get ionized as soon as the electrons bind to form an atom. But as the universe cools to about 100,000°C, the electrons are moving slowly enough not to be able to get away from the attractive force of the proton and would then form atoms (hydrogen atoms, helium atoms, etc.). The age of the universe at that time is about 400,000 years, still pretty young compared to the age of the universe now.

Then, there is an epoch called the matter radiation equality epoch, sometime after the electrons combine with protons and the helium nucleus to form atoms. At this time, dark matter will start dominating the energy density of the universe, and then structures in the universe such as galaxies will begin to form. A summary of major events in cosmic evolution is given in Table 6.1.

No elements heavier than deuterium, helium, and lithium were formed during BBN because of the nature of nuclear forces. These nuclei later became part of stars and galaxies. Of course, some of these nuclei will be destroyed during the process of star formation, but their overall abundances remain largely unchanged. Their copiousness can now be deduced from astronomical observations and compared with theoretical predictions. Whatever amount has been observed seems to fit the

Table 6.1. Timeline of cosmic evolution: major events, ages, temperatures.

What Happened?	Age of the Universe at the Time of the Event	Temperature of Cosmic Soup at the Epoch
Inflation	$\sim 10^{-38}$ seconds	$\sim 10^{16}$ GeV
Matter–anti-matter asymmetry	?	?
Electroweak phase transition	10^{-10} seconds	200 GeV
Formation of protons and neutrons from quarks	10^{-5} seconds	200 MeV
Big Bang nucleosynthesis	1 minute	Few MeV
Matter radiation equality	30,000 years	1 eV
Recombination to form atoms	300,000 years	0.33 eV
First star formation	100 million years	~10 milli-electron volts
Re-ionization	400 million years	~ 3 milli-eV
Dark energy domination	10 billion years	0.3 milli-eV
Right now	13.8 billion years	0.24 milli-eV

calculations at the time of BBN only if the number of neutrinos is slightly over three together with for one photon and one electron contributing to the expansion of the universe. No other light particles are allowed to be there for the BBN results to agree with observations. This is a very stringent constraint anytime new physics is added to the standard model of particle physics and cosmology.

The question one can always ask is: Why are the laws of our universe the way they are? This is a much deeper and much broader question that has been widely discussed at great length for many years. We do not quite know the answer, although Einstein's theory provides the basic guiding equation. There are some who invoke the so-called anthropic principle to answer this question. This principle, as stated in chapter 2 suggests that laws of nature take values that allow the existence of life capable of asking such questions. There is still a great deal of debate on this idea, which is not pursued further here. If the goal of science is to answer all questions about nature, this is certainly a valid one to ask regarding the evolution of our universe. One may hope that an answer will eventually be found, although it is possible that something like the anthropic principle may remain the only explanation.

6.6. Galaxy Formation and Role of Dark Matter

Dark matter is the reason galaxies form out of the uniformly expanding cosmic fluid, as already briefly noted in Chapter 2. The way it is believed to have happened is that during cosmic expansion, at some point, dark matter forms clumps of higher density compared to the surroundings. These clumps exert strong gravitational attraction on surrounding matter. They form what are called potential wells. Visible normal matter is then pulled into these wells to form galaxies and stars we see in the sky. This is literally like a person walking at random at night in the backyard that contains a well with an open mouth. If persons walk into this opening unknowingly, they will fall in and get trapped. Baryons essentially suffer the same fate; when large amounts of matter fall into such wells, galaxies are formed. How galaxies form is actually not completely understood, since the detailed properties of dark matter remain unknown. Nonetheless, this provides a broad picture.

Stars are believed to form from massive dust clouds coagulating over time within galaxies, and some galaxies may have grown through such processes. The details are not completely clear and are still a matter of active research. Roughly speaking, it could have happened that stars get close to each other and their gravitational attraction pulls them to form galaxies.

The first such computer study for galaxy formation using dark matter was carried out in 1997 by three astrophysicists, Julio Navarro, Carlos Frenk, and Simon White [53], who derived a profile describing how matter is distributed within galaxies. They made detailed predictions about the distribution of galaxies in the universe within a WIMP-like picture of dark matter. The predicted profile of galaxies is largely successful, but shows some discrepancies with observations. Their work leads to the conclusion that the density of matter at the center of a galaxy is very high compared to its outer regions. However, observations do not fully support this prediction. To cure this problem, David Spergel and Paul Steinhardt [54, 55] suggested that the dark matter might be exerting forces on each other rather than just flying past each other. This would prevent the dark matter from coagulating at the center and pushing each other by the self-interaction force. Such dark matter is called self-interacting dark matter. Evidence seems to be gradually accumulating in favor of such a possibility [54, 55], although it is not yet conclusively established.

Another problem that arises from the Navaro *et al.* work on structure formation is that it predicts too many dwarf galaxies in a typical galaxy like our own Milky Way. Dwarf galaxies are smaller galaxies compared to typical spiral galaxies like ours, about 10 or more times smaller with fewer stars, as noted. They are typically dominated by dark matter. In a typical dwarf galaxy, the total mass of visible stars is insufficient to explain the speeds at which the stars are moving. Of course, because they contain very few visible stars, dwarf galaxies are, in general, faint and difficult to observe. The cold dark matter model predicts more than 100 dwarf galaxies in the vicinity of the Milky Way, whereas only about 38 have been observed so far. This discrepancy is known as the missing satellite problem.

Mirror world models provide a natural resolution of the core density problem since mirror dark matter, being analogous to normal matter, can have self-interactions that prevent excessive accumulation at the galactic centers. Self-interacting dark matter may also help address the missing dwarf galaxy problem, although this possibility remains under dispute.

6.7. Mirror World: A Cold Dark Place

Now that we have an idea of the evolution of a single universe, we return to discussing what happens when there is a mirror sector.

The first obvious question that arises in any discussion of many universes is: "Are the laws governing the different universes the same as those in our universe? If they are not the same, are they similar but different only in details, or are they completely different?" Secondly, "do the laws allow the universes to evolve in the same way or differently?" In a general many-universe scenario, we do not know the answers to any of these questions. In the mirror universe case, however, mirror symmetry provides some guidance in addressing this question. The nature of forces in the mirror sector is the same at the beginning of the Big Bang due to mirror reflection symmetry of the theory; as a result, some of the discussion could be true in general. However, the details can differ, since the evolution of the mirror sector need not be the same in all respects.

It is safe to assume that in the very early moments of the Big Bang, mirror symmetry guarantees that the rate of expansion of the mirror sector is the same as that of our sector. One detail that could affect the evolution of the mirror sector is whether mirror symmetry is broken. In this case, the evolution of the universe before and after the temperature at which the symmetry was broken could be very different in the two sectors. This is very similar to the case of supersymmetric theories discussed earlier, but unlike the supersymmetry case, the mirror symmetry may not be so strongly broken. Some of the changes in laws in this case are also dictated by simple observational constraints in our universe, as discussed in the following section.

To delve further into the cosmic evolution of the mirror sector, we will begin with the constraints from what we know about our sector.

The first important implication of what we know for sure about our universe is that the mirror sector must have a much lower temperature than "our side." The particular observation that leads to this conclusion is the currently observed abundance of helium and deuterium nuclei in the universe [47, 48].

To understand the reason for this, recall the discussion of BBN from the previous section, when the helium and deuterium nuclei are synthesized out of protons and neutrons. A key step that initiates this process is when the weak process involving neutrons and protons ($n \rightarrow p + e + v$) the reverse reaction ($v + p \rightarrow e^{+} + n$) occur at a rate slower than the Hubble expansion rate. The Hubble rate, on the other hand, is determined by how many neutrinos, photons, and electrons are present at that epoch, since the more such particles there are, the greater the energy density and the faster the expansion rate of the universe. In the single-universe model, there are three standard model neutrinos, one photon, and one electron, which are sufficient to produce the observed amount of helium in the universe. The mirror universe model, on the other hand, has six neutrinos, two photons, and two electrons, as noted earlier.

What happens when there are more species of neutrinos, photons, and electron-type light particles is that the universe expands at a faster rate, since each particle carries energy and puts pressure on the universe to expand faster. In such a universe, the weak force reactions would become slower, or "freeze out," much earlier than in a single-universe model. A property of collections of atoms or tiny particles in a gas is that the more massive ones are less numerous than the lighter ones, and the extent of this difference depends on the temperature of the particle soup. Thus, the temperature at which weak interaction "freezes out" occurs affects the relative amounts of protons and neutrons. Since the amount of helium nuclei formed from the combination of protons and neutrons ($2p + 2n = He^{4}$) depends on their relative numbers, the helium abundance is affected in a mirror universe model. In fact, if nothing new were assumed about the mirror universe, there would be an equal number of protons and neutrons, leading to a universe composed entirely of helium and very little hydrogen. This is inconsistent with the observations. Therefore, for the mirror universe idea to be observationally viable, the effects of mirror neutrinos, mirror electrons, and the mirror photons at the time of BBN must be suppressed. A detailed mechanism for achieving this was proposed in Ref. [56] (see also Refs. [57, 58]).

The basic idea of Ref. [56] is called asymmetric inflation. As discussed in Section 6.2 of this Chapter, the early universe underwent a rapid phase of expansion, which is needed to understand what we observe about the universe today. During the inflationary expansion, there were no standard model particles in the universe. How then did the universe we are familiar with emerge? It turns out that, as the inflationary expansion ends, the particle (believed to be a boson like the Higgs boson) that caused the inflation begins to decay, and this decay leads to a thermal bath of standard model particles. This causes the universe to heat up, and the final temperature is called the reheating temperature of the universe.

In a two-universe scenario like ours, there will be a similar phenomenon in both sectors of the universe. Both sectors will undergo inflation, and at the end of this period, the Higgs-like field (called the inflaton field in scientific literature), responsible for inflation, will decay to both familiar visible sector particles as well as the mirror sector particles. If an equal number of particles are produced per unit volume in both sectors, their temperatures will be identical, and the BBN problem will persist. However, if the inflaton (the particle responsible for inflation) decay produces fewer mirror sector particles than the standard model particles, then the mirror sector will contain fewer neutrinos, photons, and electrons, thus resulting in a lower temperature. In this case, the predictions of BBN remain consistent with the observed abundances of helium, deuterium, and lithium. The element abundances remain the same as in the situation with no mirror world. Fewer mirror particles produced will keep the mirror sector at a lower temperature compared to the visible sector, thus motivating the description of a "cooler mirror sector."

This modification of the standard inflation theory is called asymmetric inflation [56–58]. This can be realized in a theoretically consistent way using the standard techniques of inflation model building [56]. The need for asymmetric inflation in mirror world scenarios is independent of how the two universes manifest today; in other words, regardless of whether the present-day mirror sector is symmetric or asymmetric, there has to be asymmetric inflation at the beginning for the model to be consistent with the observation of primordial helium, deuterium, and lithium. For a summary of further consequences of the cooler mirror world scenario, see Ref. [41].

Finally, to make sure that the asymmetric inflation picture serves its purpose of satisfying the BBN constraints, there must be no "strong" new interactions connecting the particles of the standard model to their mirror counterparts. Otherwise, such interactions would thermalize the two sectors and equalize their temperatures, reintroducing the BBN inconsistency.

Chapter 7

Symmetric Mirror World and Implications

"The difficulty lies not so much in developing new ideas as in escaping from old ones."

— **John Maynard Keynes**

When there are two sides to a story, the natural human impulse is to examine them, compare them, and possibly choose a side. In the discussion of the mirror universe, we have reached that point. In the previous chapter, we briefly outlined the symmetric and asymmetric mirror worlds. In this chapter, we focus on the symmetric world and discuss some of its implications and tests.

The identical couplings of particles to Higgs fields in both worlds implies that, provided the Higgs spread values are identical, all particle masses in the mirror world will be same as the corresponding particles in our world (the former will be called the mirror partners, as we noted previously). This means that the proton (p) in our world will have the same mass as its mirror counterpart, the mirror proton (p'). This conclusion holds for the other particles as well. Similarly, the strengths of the strong and weak forces will also be identical in both sectors. This is a very symmetric situation indeed, and hence, we will call this the symmetric mirror model.

Identical particle masses in both worlds imply that if the mirror protons constitute the dark matter in a symmetric mirror world, their abundance must be five times that of the visible world. This may be possible due to the lower temperature of the mirror world, which will affect the evolution of the mirror sector in a manner different from the visible sector. Other important implications of the symmetric mirror world are discussed in the following section, along with the ways to test them experimentally.

7.1. Symmetric Mirror World and Neutron–Mirror Neutron Oscillation

A neutron in our world is the particle that partners with the proton, its positively charged brother, to help build the whole world of chemical elements. These elements populate the famous periodic table, invented by Mendeleev in 1869, which adorns the wall of every chemistry department in the world. These elements and the compounds formed from them are used by us every day. Without neutrons, the heavier chemical elements essential for life cannot form. Therefore, a neutron is a crucial ingredient of life in the universe.

How much do we know about neutrons since Chadwick discovered them in May 1932? We do know a lot: We know that if a neutron is left on its own in free space, it will fragment to a proton, an electron, and an anti-neutrino due to the weak force and will disappear in about 15 minutes. So, how does it build elements that we know are stable, long-lived, and constitute the universe? It turns out that when a neutron is inside the nucleus of an element, it becomes stable. It falls into the "well of nuclear potential energy" from which it cannot climb out on its own and decay. This is much like an animal that falls into a well and cannot get out without external help. Of course, if there are too many neutrons inside a nucleus, the well mentioned above gets filled up, and the neutron at the top can easily climb out. We then say that the nucleus becomes unstable and decays. This is the basis of nuclear energy generation. Thus, there must be the right balance between neutrons and protons in a nucleus to make some atomic nuclei with neutrons live forever.

Broadly speaking, that is all we know about the neutron now; however, if there is a mirror sector in the universe, the neutron can acquire new properties. In particular, in the symmetric mirror world theory, where the neutron has an equal-mass partner, this leads to some interesting possibilities. One new possibility is that the neutron can suddenly transform into the mirror neutron if there are new weak forces allowing this to happen. This is possible because the neutron does not carry any electric charge, and the mirror neutron does not carry any mirror electric charge, and they have nearly the same mass. An electric charge will prevent a proton from undergoing this kind of oscillation to a mirror proton or an electron from transforming into a mirror electron, even though their masses are nearly the same. Since the neutrinos and photons do not carry electric charge, they can also oscillate to their mirror partners.

This new phenomenon leads to the possibility that a free neutron left to propagate in free space can oscillate on its own into the mirror neutron [59] using the new force, as mentioned above. This is much like the neutrino oscillations, which happen if there is a "mass interaction" connecting two neutrinos. This has already been observed for known neutrinos in nature and was discussed in Chapter 3. There are also other elementary particles, such as a K-meson, that oscillate into their anti-particle in the same way. In the case of the K-meson, the forces connecting the K-meson to the anti-K-meson arise from the familiar weak interaction processes, as suggested by Murray Gell-Mann and Abraham Pais in 1955. This process was already observed many years ago. In 1980, it was pointed out that in a theoretical model that unifies weak, electromagnetic, and strong forces, the neutrons could also oscillate to their anti-particle, the anti-neutrons [60], and several experiments followed soon afterward looking for it. It has not yet been discovered. The new possibility of a neutron oscillating to a mirror neutron [59] has caught the imagination of experimentalists in recent years, and there are now several experiments searching for it [61].

In the neutrino oscillation case, the masses are so tiny that they need not be identical to oscillate to one another. However, since the neutron is more than a billion times heavier than the neutrino, the masses of n and n' must be nearly the same for oscillations to be observable, which is the case in the symmetric mirror model.

The n–n' oscillation is a process that is being actively explored at various reactor neutron facilities in the US and Europe [61], including the Oak Ridge Neutron Facility in Tennessee, the PSI laboratory in Zurich, Switzerland, and the European Spallation Source in Sweden. Its discovery will be transformational and will break new ground in both particle physics and cosmology. Observation of this effect would directly establish the existence of a mirror sector to the universe and would be of significant importance for dark matter physics.

One important point about *n–n′* oscillation is that once the neutron is inside a nucleus, the oscillation is forbidden. Therefore, the search must be conducted in as pure a vacuum as possible using free neutrons.

The way to search for *n–n′* oscillation is to use a neutron beam in a reactor, allow the neutrons to freely fly a distance, and then finally see how many neutrons have disappeared from the beam. Currently, experiments are sensitive to oscillation times of the order of a few seconds. So far, no such effect has been noted, implying that the oscillation time must be longer than about 10–100 seconds or so. More sensitive experiments are currently in the planning stage to improve this bound.

Another way to search for *n–n′* oscillation is to use a neutron bottle that stores ultra-cold neutrons and watch how many neutrons disappear over time. This method is being used in an experiment at the Paul Scherrer Institute in Switzerland.

An interesting question that now emerges is how the different nuclei remain stable if a neutron becomes a mirror neutron. The answer is that when protons and neutrons bind to form a nucleus, their total mass is less than the sum of their original masses due to the binding force. The effective neutron mass goes down inside a nucleus. Since a mirror neutron does not bind to a nucleus, its effective mass does not go down. Thus, the conversion of a neutron inside the nucleus into a mirror neutron would produce a free particle and lead to an energy increase between the initial and final states. This mismatch will violate the conservation of energy, a sacred principle of physics. Thus, the stability of nuclei is not affected by *n–n′* oscillation.

A second important point is that if a mirror neutron is in a medium, the oscillation is again suppressed, implying that a mirror neutron, once produced, can fly through an opaque medium made of normal matter.

This n–n' oscillation process, if it exists, has several other important consequences, which will be discussed in the following section.

7.2. Neutrons Shining Through the Wall

A normal beam of neutrons, when incident on a thick wall, would not be able to penetrate it, since the neutrons, because of their nuclear force, would be absorbed by the atomic nuclei on the wall and stopped from going very far. If the neutrons are too energetic, very few may, of course, be able to penetrate through some materials. The situation becomes completely different if neutron–mirror neutron oscillation takes place. In this case, some of the neutrons can oscillate into mirror neutrons before reaching the wall. Since a mirror neutron does not interact with the normal nuclei in the wall material and its oscillation back to a neutron is suppressed, it will pass through the wall unimpeded and reach the other side of the wall. After the mirror neutron reaches the other side of the wall, it can oscillate back to a normal neutron, since outside the wall is essentially free space. The final neutron can be detected in a neutron detector. The net effect is that due to neutron–mirror neutron oscillation, the neutron could pass through the wall, which normally it could not have [62]. This is called shining through the wall and can provide a direct confirmation of the phenomenon of neutron–mirror neutron oscillation. Clearly, neutrons shining through the wall, if confirmed, will have many practical applications (Fig. 7.1).

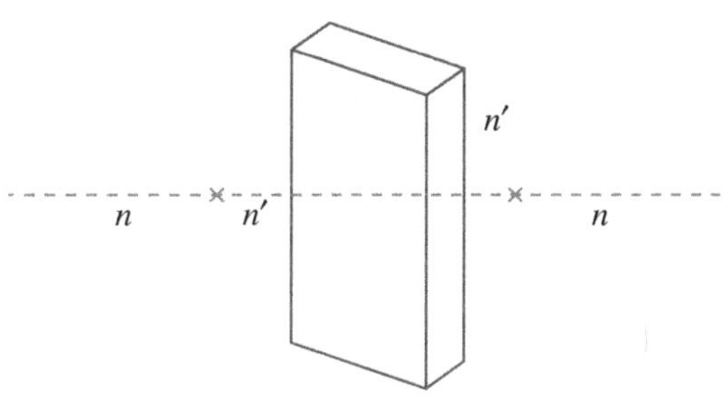

Fig. 7.1. Cartoon of neutrons shining through the wall. The mirror neutron passes through the wall unscathed and appears on the other side before converting to a normal neutron.

7.3. Neutrons Rising from the Earth

Another spectacular way in which the neutron–mirror neutron oscillation can manifest is through neutrons rising from the Earth and going up during a solar flare. The rising neutrons can be detected by a neutron detector in a satellite orbiting the Earth. The reason that can happen is as follows.

The Sun emits flares of energetic particles, often during the periods when Sunspots appear every 11 years. In a solar flare, energetic neutrons with energies of 1–10 million electron volts are emitted along with other particles. Such solar-flare neutrons have been detected by the NASA MESSENGER spacecraft flying near planet Mercury in 2014. It is these neutrons that are of interest to us. Most of the neutrons, of course, decay before reaching the Earth if they are not energetic enough, but some may survive. The surviving neutrons on their way from the Sun to the Earth could convert to mirror neutrons before they penetrate the Earth's surface. Once a fraction of them turn into mirror neutrons, the latter would sail through the interior of the Earth as if there were no barrier and come out on the other side from the ground as mirror neutrons. As the emerging mirror neutrons keep going up, they can convert back to normal neutrons, which would then be detected in the neutron detector flying above the ground (Fig. 7.2).

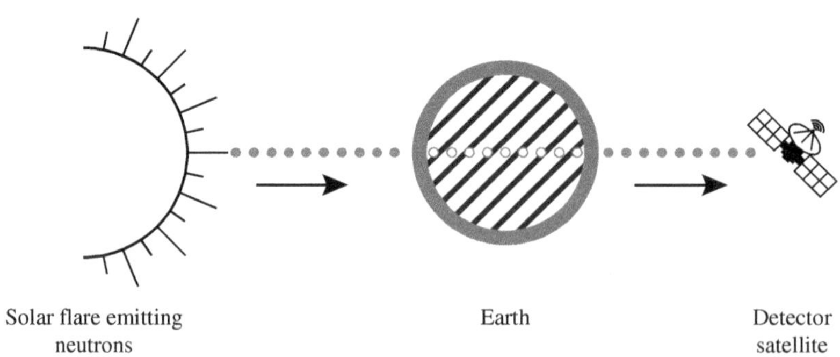

Fig. 7.2. Illustration of solar flare neutrons rising from the Earth. Solid circles signify the normal neutrons and hollow circles the mirror neutrons. Inside the Earth, the presence of matter prevents the mirror neutrons (hollow circles) from converting back into neutrons, as in the case of "shining through the wall" neutrons.

The consequences of n–n' oscillation have profound implications for the neutron stars that we see in our sector of the universe, and these will be discussed in the following section.

7.4. Possible Applications

One possible application of the neutron–mirror neutron transformation is imaging the interior of the Earth. The idea would be to shine a high beam of neutrons on the Earth, some of which, before hitting the Earth, can become mirror neutrons. The mirror neutrons then travel unimpeded through the Earth (since they have no interaction with familiar matter) and emerge on the other side, where they can reconvert back into familiar neutrons. As the mirror neutrons travel through the Earth, their energy pattern would be sensitive to the density of normal matter they pass through. This pattern of energy modulation has been observed for neutrinos and will be transferred back to the neutrons when the mirror neutrons convert back. Thus, much like ultrasound waves revealing the interior of any object it passes through or seismic waves revealing the shape of magma chambers below the Earth's crust, n–n' conversion could reveal the interior of the Earth. This could be useful in applications ranging from earthquake prediction to the detection of oil and mineral deposits underground.

Neutron–mirror neutron oscillation could also be useful in detecting clandestine nuclear explosions below the Earth's surface. Since large amounts of neutrons are emitted in a nuclear explosion, some of them could convert into mirror neutrons, emerge at the Earth's surface, and then reconvert into normal neutrons. These could then be detected and provide an indication of a nuclear explosion.

Chapter 8
Asymmetric Mirror World

> "Do not go where the path may lead, go instead where there is no path and leave a trail".
>
> — **Ralph Waldo Emerson**

The asymmetric mirror world is the other side of the mirror world story. The idea is as unique and intriguing as that of the symmetric mirror world. As noted in Chapter 6, in the asymmetric mirror scenario, the masses of the mirror sector particles are different from those of the corresponding particles in the visible sector of the universe. This difference could lead to various kinds of new phenomena that we discuss here. One possibility is that our neutron could decay into a mirror neutron if the latter is lighter than our neutron. This possibility has been invoked to explain certain apparent anomalies in the decay properties of the neutron, which we discuss in the following section.

8.1. How Long Do Neutrons Live?

As already stated, a neutron is one of the fundamental building blocks of matter and life, along with the proton and the electron. Protons and neutrons make up the nucleus of all the various atoms; the planets and stars in the universe, as well as the chemicals in our bodies, all contain neutrons. Neutrons thus play a vital role in the universe. The absence of electric charge on neutrons helps bind them to protons inside the nucleus. Neutrons are also important for energy generation and in medicine, such as neutron therapy for cancer. It is therefore very important to know as much as possible about neutrons and to be as precise as possible in our knowledge of them.

There are many research programs involving neutrons around the world. They use neutrons generated from various nuclear processes. The two major sources of neutrons are as follows: (i) nuclear reactors, where the neutrons are generated by the fission of uranium nuclei; (ii) the spallation neutron sources (SNS), where a beam of protons impinges on a collection of nuclei (e.g., mercury) to produce beams of neutrons through nuclear disintegration.

Neutrons, unlike protons and electrons, are not stable particles. Left to themselves, they decay and disappear in about 15 minutes. They decay into a proton, an electron, and an anti-neutrino. This process takes so long because the forces responsible for this decay are very weak forces (for a detailed discussion on this, see Chapter 3). Once bound inside nuclei, however, the neutrons remain stable. This is how the periodic table of elements is built up.

One of the important questions about neutrons is how long they live when they are free. The neutron lifetime is a key input in understanding the formation of elements in the evolution of the universe, including the Big Bang nucleosynthesis discussed in Chapter 5, and the formation of heavy elements in astrophysical environments, such as supernovae or neutron star mergers.

The measurement of the lifetime of the neutron is carried out in two kinds of experiments: one is called the bottle experiment [63], and the other is called the beam experiment [64]. In the bottle experiment, the slow-moving ultra-cold neutrons are collected in a bottle, and their number is counted at the beginning as well as after certain time intervals. Using this information, physicists can determine how long neutrons live. In the beam experiment, the initial number of neutrons is counted, and the neutrons are then allowed to flow in a beam. At the end of the beam, the number of protons produced from their decay is counted [65].

Normally, one would expect the same value for the lifetime from both experiments. However, despite repeated efforts with both methods, the beam experiment gives a lifetime that is about nine seconds longer than that obtained from the bottle experiments (see, however, Ref. [65]). How could that happen? The nine seconds may appear to be a small number, but the precision of the experiments and the theory behind the calculation of the lifetime are so good that the discrepancy is believed to be an indication of new fundamental physics rather than an experimental error. This has given rise to theoretical speculation, some of which fit quite well into the mirror world model. In particular, there are two speculations of interest [66, 67].

The first possibility is that the neutron in the beam decays part of the time into a dark particle rather than into a proton, an electron, and an anti-neutrino. Since in the beam experiment, only the protons from neutron decay are counted, they will miss these dark particles if there is a final state that involves a dark particle. Therefore, the rate of neutron decay will be measured to be smaller, and the lifetime measured will appear longer. In contrast, in the bottle experiment, the final-state protons are not counted, and one would measure the true lifetime. According to this picture, the bottle experiment measures the true lifetime, while the beam experiment does not. The connection to the mirror world is that the new dark final state of the neutron could involve a mirror particle, which will not be observed [66] in the beam experiment. For example, this could be a lighter mirror neutron and the mirror photon. This will explain why the beam measurement gives a longer lifetime.

In the second explanation, one uses the neutron–mirror neutron oscillation as a way to explain the discrepancy [67]. In the beam experiment, due to the large magnetic field applied, a small fraction of the neutrons can oscillate to mirror neutrons, so that their decay will not contribute to the measurement of the neutron lifetime [67]. Thus, the mirror model provides two ways to explain the observed neutron lifetime anomaly.

8.2. Why Are the Matter and Dark Matter Contents of the Universe Similar?

One curious fact about the amount of dark matter in the universe is that it differs from the matter excess in the universe by only a factor of five. Since matter and dark matter are normally expected to be different kinds of particles with different forces acting on them, this fact adds another puzzle to the already existing dark matter mysteries. It is not easy to understand why this is so. This is called the matter–dark matter coincidence puzzle.

Of course, if dark matter is a mirror partner of particles in our world, this similarity becomes less mysterious. However, in the absence of a detailed theory, it is still not completely clear why this happens. If we look more closely at the detailed behavior of the two-world picture, plausible arguments emerge that can explain this similarity. This would require an asymmetric mirror world picture, as we explain in the following.

As discussed earlier, in the universe we live in, there is more matter than anti-matter. We do not yet know how this asymmetry arises, although many scenarios have been proposed over the years to explain it. It is therefore not too much of a stretch to imagine that a similar process might occur in the mirror sector, leading to the production of an excess of mirror matter over mirror anti-matter. In fact, the symmetry between our world and the mirror world would naturally suggest such a possibility. This process would then produce more mass in a mirror hydrogen or a neutron over its anti-particle, and these could constitute the dark matter of the universe. Since the production processes of excess matter and excess dark matter over their anti-matter are similar, this line of reasoning would suggest that the amounts per cubic centimeter of matter and mirror matter are the same. In fact, there are some theories that predict such a possibility [42, 68, 69]. In that case, if somehow the mirror proton or neutron is five times heavier than our proton, as could be the case in an asymmetric mirror world, it would resolve a deep mystery of dark matter physics and explain why the dark matter mass content of the universe is observed to be only five times more than the visible mass content in our universe. Thus, the matter–dark matter coincidence puzzle might just be explained in a mirror world picture, adding another feather in the cap of the mirror world explanation of dark matter. Such a scenario is more likely if dark matter arises in a way similar to the generation of the visible matter–anti-matter asymmetry, the so-called asymmetric dark matter picture [70].

8.3. Are Neutrinos' New Cousins Showing Up?

We saw in Chapter 3 that the standard model of particles and forces contains three neutrinos accompanying the quarks and the charged leptons, electron-type, muon-type, and tau-type neutrinos, named after the charged lepton they accompany. The electron neutrino participates in the observed weak interactions such as beta decay and hydrogen fusion at the core of the Sun. In the beta decay process, the anti-neutrino is always accompanied by the electron. In many similar weak force-mediated processes, the other neutrinos also appear both in nature and in the laboratory, in the company of the corresponding charged leptons: muon or tau lepton. For example, the muon neutrinos are produced in cosmic rays together with muons. The tau neutrino is produced mainly in accelerators, accompanied by the tau lepton. All of these neutrinos have been observed and together help explain the neutrino oscillation phenomenon observed for neutrinos coming from the Sun, the atmosphere, and particle accelerators. As already explained, oscillation phenomena have established that all these neutrinos have mass, which allows them to transform from one type into another. The oscillations also determine the differences between the masses of the different neutrino species, since the frequency of the observed oscillations depends on these mass differences. From the study of oscillations, we now know that the neutrino masses are all smaller than about a tenth of an electron volt, i.e., less than a tenth of a million times smaller than the mass of the electron.

Neutrinos are fascinating particles with a much loftier role in the universe than their petite profile would suggest. For example, they are responsible for building up the periodic table of chemical elements, which is the basis of human life. It is therefore more than mere curiosity to ask whether there are any other kinds of neutrinos that could possibly teach us more fascinating things.

During the last 30 years, several experimental results have been announced from different oscillation experiments, which cannot be explained within the conventional three-neutrino framework. They have therefore led to speculation that there may be extra neutrinos beyond the three present in the standard model. These extra neutrinos are called sterile neutrinos because they do not directly participate in the known weak interactions, such as beta decay. In other words, they are sterile or inert with respect to known weak interactions. In fact, their interactions with the weak force-mediating bosons W and Z are forbidden or very highly suppressed.

We had some discussion of such neutrinos in Chapter 4 in connection with dark matter and noted how if one of them has a mass about a few hundred times lighter than the electron, it can be a dark matter of the universe. Our discussion here will also refer to similar particles, but with masses about a thousand times smaller than the above-mentioned dark matter neutrinos. In the following section, we summarize the experiments that have provided hints of such particles.

8.3.1. *Why Introduce Sterile Neutrinos?*

To understand the need for introducing sterile neutrinos, a brief explanation of the connection between neutrino oscillation and the mass of the neutrino is needed. In neutrino oscillations, one defines an oscillation length, which roughly measures the distance over which one type of neutrino changes from one kind to another while traveling through empty space. The oscillation length is determined by the mass of the heavier of the two neutrino species involved and by its energy. If the oscillation length is very large, either the mass of the heavier neutrino participating in the oscillation is small or the energy of the oscillating neutrino is high. The reverse is true for shorter oscillation lengths. For example, in the oscillation of the neutrinos from the atmosphere, the oscillation length is a few kilometers. Since their energy is known, they give the tau neutrino mass to be roughly a 20th of an electron volt (one electron volt being roughly one millionth of the mass of an electron). For oscillation of neutrinos from the Sun, the situation is more complicated because the neutrino travels a long distance through the dense matter inside the Sun and experiences diffraction much like light travelling through a prism [71]. This modifies the oscillation length in such a way that the corresponding mass is about a hundredth of an electron volt, even though the distance between the Earth and the Sun is much larger.

Now, let us look at the new kinds of evidence for oscillations besides the solar and atmospheric oscillations. In 1993, an experiment carried out at Los Alamos Laboratory in New Mexico showed that a muon neutrino born in the decay of the pi meson transforms to an electron neutrino after travelling a much shorter distance. Since the energy of these neutrinos is known, the short distance implies that the corresponding mass of the heavier neutrino is about 1 electron volt. The mass of this neutrino turned out to be much larger (about 20 times the tau neutrino mass) than the masses of the known neutrinos, i.e., those participating in either solar or atmospheric oscillations. So, the question that then arose was: What kind of neutrino is this, or is this phenomenon not related to neutrino oscillation at all? No mechanism other than neutrino oscillation has been able to explain the Los Alamos result. This experiment is known as LSND (short form for Liquid Scintillator Neutrino Detector). The simplest interpretation of this result is that there exists a new neutrino beyond the three standard model neutrinos. This result was so intriguing and had such groundbreaking implications that many other experiments followed, and several are still underway to verify it [72, 73].

An experiment in the UK at the Rutherford laboratory in Chilton, called KARMEN, was carried out immediately after the Los Alamos experiment to check the LSND results. It did not find any positive evidence confirming the LSND results, but it could not fully rule it out either. This was followed by another experiment called Mini-BooNE, performed at Fermilab near Chicago in 2002 to test the LSND claim. Its results were controversial, and again, they did not contradict the LSND findings. These results are now being more thoroughly examined in a new Fermilab experiment called MicroBooNE, which uses liquid argon as the detecting material. Liquid argon is useful because of its easy availability and excellent particle-detection properties.

Similar indications for sterile neutrinos also came from another experiment involving neutrinos from a chromium source. These neutrinos were used for calibration of the gallium nucleus used in the search for solar neutrinos. The goal was to understand how neutrinos scatter against the gallium source, the knowledge of which would help interpret measurements of neutrinos coming from the Sun. Since the same type of neutrinos is involved, one type of scattering experiment can help in understanding another. These are called radioactive source experiments. One result of the chromium-source neutrino experiment was that some of the emitted neutrinos seemed to disappear within a few meters, in a manner similar to what was observed in the LSND experiment.

One interpretation of this loss is that the missing neutrinos oscillated into another species of neutrinos. Again, as before, from the known distance and energy of neutrinos, one could conclude that this was an electron-volt mass neutrino, similar to that in the LSND case [72, 73]. All these results need to be confirmed, but they have stimulated a rapid surge of research aimed at understanding such a new species of neutrinos. If they were confirmed to exist, this is where mirror universe models could become relevant.

As explained earlier, these new species of neutrinos are called sterile neutrinos because they are not emitted by gauge bosons like the *W*- and *Z*-bosons of the standard model, unlike the three familiar neutrino types discussed in Chapter 2. Since these experiments so far have been inconclusive, the existence of sterile neutrinos remains an intriguing but unconfirmed signal of new physics.

8.4. Mirror Universe and Sterile Neutrinos

As discussed earlier, mirror world models involve a duplication of all particles in the standard model. Thus, there is a triplet of new neutrinos in the mirror sector, each being a mirror partner of the corresponding standard model neutrino. They do not participate in the known weak interaction because they do not interact with the familiar *W*- or *Z*-bosons, but rather with their mirror partners, the W'- and Z'-bosons. Thus, these new mirror neutrinos are sterile with respect to our weak interactions. If their masses can be made higher than those of the familiar neutrinos, they could be strong candidates for the sterile neutrinos discovered in the LSND experiment.

Of course, in the symmetric mirror model, the masses of all mirror fermions are the same as those of the visible fermions; therefore, a sterile neutrino with an electron volt is difficult to accommodate in the symmetric mirror model. On the other hand, in the asymmetric mirror model, mass differences between known particles and their mirror particles arise more naturally. Thus, if a sterile neutrino such as the one indicated by the LSND experiment is confirmed, it will provide strong support for the asymmetric mirror model [74]. A theoretical possibility is that one or two of the mirror neutrinos have masses in the electron-volt range, while the third one has a higher mass, making it a possible candidate for warm dark matter.

8.5. Mirror Neutrino Effect on the Cosmic Neutrino Background

If the LSND results are confirmed by other experiments searching for sterile neutrinos, a natural interpretation is that they are mirror neutrinos that can transform into the known neutrinos. That would have important implications for the cosmic neutrino background (CνB).

What is the CνB? Just like there exists a cosmic microwave background (CMB), which is the remnant radiation from the Big Bang, a relic neutrino background filling the universe would be expected from the Big Bang expansion of the universe. The prediction is that, if neutrinos are stable, there will be about 330 of them per cubic centimeter, and they will form the CνB.

The CMB has of course been studied in extreme detail in many experiments, such as COBE (cosmic background explorer), WMAP (Wilkinson microwave anisotropy probe), and the PLANCK observatory (operated from 2009 to 2013 by the European Space Agency). These studies have provided us with fundamental insights into the early universe. The CνB, which is believed to exist all around us, has, however, not been studied because neutrinos interact so weakly, making it difficult to measure and study them. However, as the efficacy of scientific instruments improves, it will eventually become possible to study them. Such advanced measurements will give us a full picture of how many neutrinos exist at a given energy, just as we have for photons through detailed observations of the CMB.

The existence of heavier mirror neutrinos that mix with our neutrinos will allow the former to decay into the latter. These mirror neutrinos could be remnants from the early moments of the Big Bang, or they could originate from stellar evolution in the mirror sector, such as when mirror stars undergo mirror supernovae. In either case, their decay to our kind of neutrinos will leave an imprint on the CνB of our neutrinos. At present, there are a total of about 330 neutrinos per cubic centimeter in the universe, and they are all very slow-moving. The decay of mirror neutrinos into familiar neutrinos will deposit more energetic neutrinos on top of the existing CνB. As a result, once the CνB is measured, it may provide another glimpse into the mirror universe and help prove or disprove its existence. It must be noted that there are plans to measure the CνB [75].

Part III

Chapter 9

Big Picture Implications of the Mirror Universe

> "Without leaps of imagination, or dreaming, we lose the excitement of possibilities. Dreaming, after all, is a form of planning."
>
> — **Gloria Steinem**

That there could exist another universe alongside the one we observe is a drastic new concept. If true, it would surely carry many new, far-reaching implications that could affect not only our knowledge and understanding of the visible universe but also potentially our own existence in unexpected ways. As many surprising details about the familiar universe continue to emerge from modern space missions such as the famed James Webb Space Telescope (JWST) and others, including the breathtaking journeys of Voyager 1 and 2 into interstellar space, the possibility of a mirror universe promises to add more surprises for future explorations. There could be new hints that would possibly come about with the help of these probes or new ones. For physicists, astronomers, and even biologists, this presents a dream scenario and a compelling and far-reaching vision of what may lie ahead.

At a broad level, the existence of a replicate universe has the potential to add a rich diversity to the cosmos, starting from new kinds of mirror helium-rich mirror stars, mirror matter planets, and mirror galaxies to visible effects in our universe, such as overheated neutron stars and binary stars with missing partners. It is even possible that some of the new objects are residing in the vast empty voids that astronomers have seen in the intergalactic space, or they could be floating right near our galactic neighborhood, affecting the orbits of stars and planets. A whole new set of possibilities could be awaiting us as we explore the mirror universe in full glory. They may even resolve some of the puzzles in our current understanding of the visible universe. Beginning with this chapter, we delve into some of these intriguing questions.

9.1. Dark Matter in the Center of Galaxies

One long-standing puzzle that has been a nagging issue for astronomers concerns how matter is distributed in galaxies. As discussed in Chapter 6, the work of Julio Navarro and collaborators [53] provided a theoretical analysis of the implications of dark matter for galaxy formation. Among their many conclusions, one important result was the determination of a profile of how dark matter is distributed inside galaxies. They predicted that if dark matter consists of freely floating particles that do not exert forces on one another, the center of any galaxy would develop a large spike in matter density compared to the outer parts. The current observations by astronomers, however, do not seem to agree with this. They derive the distribution of matter in a galaxy by looking at the motion of stars. What they have found is that, near the center in most galaxies, the density of matter, both visible and dark, has a flat distribution without any spikes. This has become known as the core–cusp problem in astronomy. To cure this problem, David Spergel and Paul Steinhardt [54] suggested that the dark matter might be exerting forces on each other rather than just flying past without interacting significantly. This mutual dark matter force would prevent them from coagulating at the galactic center to form a cusp. The particles would push each other away by the self-interaction force as they get closer. Such dark matter is called self-interacting dark matter. Slowly, evidence seems to be building in favor of such a possibility [55], although physicists are not 100% sure about it.

In the mirror universe scenario, such self-interacting forces exist by definition. This is because the mirror dark sector is a replica of the visible side of the universe. The particles of the mirror sector, such as the mirror neutron, n', or mirror hydrogen atom, $p'e'$, which are the likely dark matter candidates in this picture, are like the neutron or hydrogen atom we are familiar with in our universe. Since we know the latter to exert forces on each other — nuclear force in the first case and residual electric forces in the second — we would expect the particles in the mirror sector to do the same. Thus, mirror symmetry could automatically prevent cusp-like galactic centers in agreement with observations.

9.2. Evolution of Structure in the Mirror Universe

Naively, one may think that in the symmetric mirror world model, the two sectors of the universe would be very similar, with similar planets, stars, plants, and animals populating both sectors. However, the real situation could not be more different for the following reason. To understand the success of Big Bang nucleosynthesis (BBN) in the visible sector, the mirror sector must be a few times cooler than ours. This can be achieved through an asymmetric inflation scenario outlined in Chapter 5 or by some other means. As we analyze the implications of this lower temperature, we see that it makes a big difference to the evolution of mirror stars as the mirror sector evolves. For purposes of simplicity, suppose the mirror sector has one-third the temperature of the visible sector. Then, at the epoch, when atoms form in the mirror sector, the electrons and protons in the visible sector are much too hot to form atoms. The formation of stars, which comes only after atoms form, will therefore take place at different times in the two sectors. Similarly, quarks bind to form protons and neutrons in the two worlds at different epochs. For example, in the mirror sector, the proton (p') forms much earlier, long before the normal protons form. Thus, without getting into too much technical detail, we note that the evolution of the visible and mirror worlds, despite their identical masses and properties, would take very different paths as the universe evolves, precisely because of the lower temperature of the mirror sector.

Broadly speaking, therefore, we do not necessarily expect mirror Earth and mirror planets to be sitting next to us due to evolutionary differences, but they could be nearby, affecting the motion of stars and planets in our sector.

The picture may be somewhat different in the asymmetric mirror world, which is defined by the Higgs spread values in both worlds being unequal. The value of the Higgs spread in our world (H) is known from laboratory measurements, such as the mass of the *W*-boson. But we do not know the value of the mirror sector Higgs spread (H′). It could be more, or it could be less than the Higgs spread value in our sector.

In our world, it is roughly 174 times the proton mass. Due to this difference, the masses of particles in the mirror sector will be proportionately different. To illustrate the effects of this, let us take the mirror particles to be a few times heavier than the corresponding particles in our sector. As we argued earlier, to get the dark matter mass content of the universe to be five times the mass content of the visible sector, it is preferable for the mirror sector particle masses to be five times heavier. This means that the proton, p', and neutron, n', can be five times heavier. For definiteness, let us choose the mirror sector temperature to be three times cooler than that of the visible sector, as before. This has several consequences [41]:

(i) The temperature at which the mirror protons and neutrons form is a few times higher than that at which the familiar p and n form. When they form, they are already moving very slowly, but their number is smaller, so they do not dominate the energy of the universe right away.

(ii) As the universe cools, there comes a time when mirror stars form. It turns out that mirror stars undergo their evolution to the end stage of their lives much more quickly, since they are much heavier. Thus, we expect that mirror stars would end up as mirror neutron stars or mirror black holes much quicker than in our sector. Such dead stars will be floating around and could give signals in a class of experiments called microlensing experiments. In a microlensing experiment, light from a star located behind a mirror star is observed. It undergoes bending due to the gravitational effects of the mirror star. The light from various bent rays can meet at one point on the other side and brighten up the spot. Thus, if the source star is moving behind the mirror star, the observed light would show a pattern of brightening, as shown in Fig. 9.1 (right panel). Note that our light can pass straight through the mirror star, as shown. This could change a little depending on whether the mirror photon mixes with the normal photon. A schematic illustration of the microlensing setup is shown in Fig. 9.1.

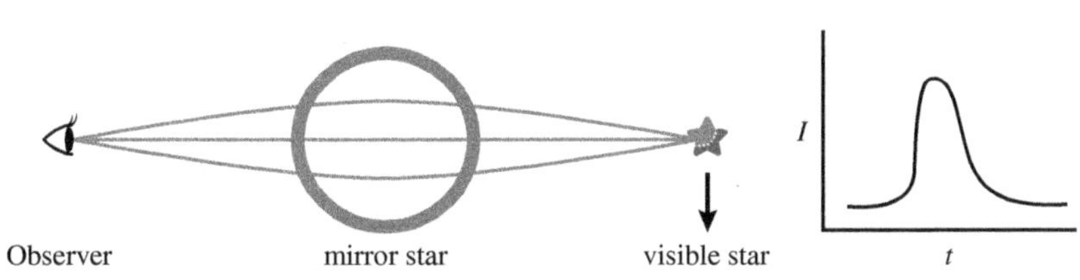

Fig. 9.1. Schematic illustration of a microlensing setup. Light from a background visible star is bent by the gravitational field of the intervening mirror star and reaches the observer along multiple paths. The panel on the right shows the resulting variation of the observed intensity as a function of time.

The right panel in Fig. 9.1 illustrates how the intensity of the starlight varies as the star moves behind the mirror star. The variation will not depend on the color of the emitted light. This phenomenon is called microlensing. Astronomers are actively conducting several microlensing searches. Experiments such as OGLE, GAIA, and Zwicky Transient Facility are constantly on the lookout for such events and have already detected thousands of them. Whether some of these events could be attributed to mirror objects remains an open question.

9.3. Mirror Helium Stars

In this section, as a second point, we argue that stars in the mirror sector of the universe are likely to be dominated by mirror helium in contrast with the visible sector, where the stars are predominantly composed of hydrogen. This property is likely to be a generic feature, common to both the symmetric and asymmetric scenarios described above.

Again, the major difference between our sector and the mirror sector of the universe in both scenarios is that the mirror sector must be colder than our sector, as discussed earlier. The coldness implies that the number of mirror photons per unit volume is much lower than in the visible sector. From the well-understood framework of BBN, we know that the amount of helium is proportional to the baryon-to-photon ratio. Therefore, unless the number of protons per unit volume in the mirror sector is extremely small, the helium content of the mirror universe could be close to 100%. What can we say about the proton or neutron density per unit volume in the mirror universe? This turns out to depend on the asymmetry between the mirror matter and anti-matter. Most theoretical analyses suggest that the mirror baryon density is either the same or somewhat larger than that in the visible sector, implying the helium-rich nature of the mirror universe.

In the asymmetric scenario, the amount of mirror helium also depends on the mass difference between the mirror proton and the mirror neutron, in addition to the density of mirror baryons at that epoch. Overall, however, the amount of mirror helium is likely to be higher in the asymmetric mirror model as well. This has profound implications for the evolution of mirror stars: They would form predominantly from mirror helium and therefore evolve much faster than the stars in our sector, which are primarily composed of hydrogen. They would burn out faster, leading to a larger population of mirror neutron stars, mirror black holes, and similar compact objects. Mirror planets, likewise, would be more helium-dominated.

9.4. Are Some Exoplanets Mirror Planets?

Over the ages, many thinkers, scientists, and curious people have asked themselves the question, "Is there anyone else out there?" In the last century, famed nuclear physicist and Nobel Prize winner, Enrico Fermi gave an answer to the above question with a contrarian point of view: "Anybody out there with a bit of rocket technology and an 'immodest' amount of imperial ambition could have conquered the entire galaxy, given that the Earth has been around for four and a half billion years." He implied that the fact that nobody has done so probably means that there is no life outside the solar system. From an objective viewpoint, however, given the sheer number of stars in the universe, how is it possible that only the solar system has life and there is no life anywhere else? This is called the Fermi paradox. This has encouraged people to speculate about life outside the solar system. It is possible to give arguments that could have prevented the Fermi scenario from happening, such as advanced civilizations outside could be recent, or space travel over such long distances could simply be very difficult. So, how do we explore whether there is indeed life (of some kind) outside the solar system?

If lessons about life on our planet are any guide, the first requisite for answering this question is to search for a planet similar to Earth at a suitable distance from a Sun-like star, with the possible presence of water. The serious search for such planets (called exoplanets since they are outside the solar system) began with the Kepler satellite launched by NASA in 2009. Kepler kept making observations till 2018, after which another satellite called the Transiting Exoplanet Survey Satellite (TESS) was launched. The Hubble Space Telescope as well as ground-based telescopes such as the CARMENES telescope in Spain, and others, have also provided valuable information about exoplanets. The James Webb Space Telescope (JWST) is expected to provide much more information.

Using the Kepler telescope, astronomers looked for periodic variations of light coming from myriads of stars and discovered many exoplanets. So far, more than 5,000 exoplanets have been discovered from a study of about half a million stars. There is a class of stars called M-dwarf stars, which have proven to be productive targets in this search. M-dwarfs are smaller than the Sun, live very long, and are particularly numerous. Their typical lifetime is 100 times that of the Sun. They have been found to have planets in the so-called habitable zone for life to exist.

A question of interest to us here is: "Could some of these exoplanets be mirror planets going around a normal star in our sector?" How can we know that? Also, since a planet going around a star is governed only by gravitational forces, there could be other possibilities, such as a mirror star with visible sector planets and a mirror star with mirror planets. Let us focus only on the first possibility.

To explore this, let us look at the various techniques used to discover exoplanets. They fall into the following classes: (i) the wobble method, (ii) the transit method, and (iii) the gravitational microlensing method, along with a few others.

The wobble method uses the fact that as the mother star pulls the planet into orbit, the planet, in turn, also gives a little nudge to the star and makes it wobble back and forth. This back-and-forth motion of the mother star can be detected using the Doppler shift of light during the wobble to infer the existence of one or more orbiting planets.

The transit method uses the fact that when a planet passes in front of the star, its light gets partially blocked and becomes dimmer. This periodic dimming of starlight indicates that there is a planet around the star.

The microlensing method uses gravitational lensing and has been explained earlier in Section 9.2. In the microlensing process [76], one star, called the lensing star, passes in front of the source star. The light from the source star is then gravitationally lensed by the lensing star and brightens up, as shown in Fig. 9.2. If there is a planet associated with the lensing star, there is a small secondary bright bump corresponding to the planet, as shown in the right panel of Fig. 9.2.

If it is a normal planet of our sector, the peak will be less bright than if it is a mirror planet, since light can pass through the mirror exoplanet as well as be bent around it. Also, the transit method will not show any change in brightness of the mother star if the exoplanet is a mirror star. Thus, by combining the transit and microlensing methods, one can identify a mirror exoplanet as opposed to a normal one.

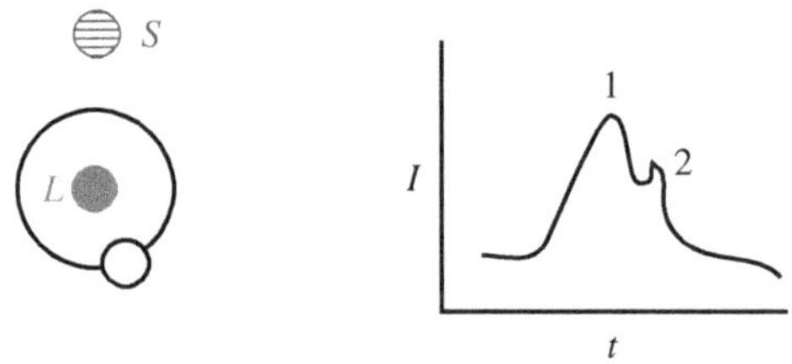

Fig. 9.2. Microlensing light curve for exoplanet search. I (vertical axis) denotes the intensity of the starlight from the source star, t denotes time, S denotes the source star, and L denotes the lensing star. The first primary bump is due to the parent lens star, while the second bump is due to secondary lensing by the planet orbiting the star as the source star S passes by.

There are some other methods used as well, such as the so-called direct imaging method, where the overwhelming glare of the star is removed, enabling the planet to be seen. This method, however, is unlikely to help us in identifying mirror exoplanets.

One curious piece of information in this regard is the discovery, about 30 years ago, of exoplanets around the star 51 Pegasi [77], one of which (HD209458b) has a mass of order Jeans' mass [78] (the Jeans mass gives the mass of a gas cloud when it begins to collapse under gravitational attraction to form a star) and an average distance of about 0.05 AU from the mother star (one AU is the distance of Earth from the Sun), some 20 times closer than Earth is to the Sun. The puzzle is that planet formation at such a close distance to the mother star is unlikely [79] due to extreme heat from the star. Note that the closest planet to the Sun is Mercury, which is about 0.4 AU, about 40% of Earth's distance from the Sun. This has led to speculation [80] that this planet could likely contain mirror matter as a huge fraction of its mass, which would not be affected by the temperature of the mother star, making its formation possible. If more such exoplanets are discovered, such speculations will gain credibility and may strengthen the case for a mirror universe.

9.5. Implications of Mirror Universe for Neutron Stars

Soon after the discovery of the neutron by James Chadwick in May 1932, Walter Baade and Fritz Zwicky in 1933 proposed the idea that there could be a star made only (or mostly) of neutrons. It would be a balance between the gravitational force and the quantum mechanical repulsion force of the neutrons, which arises due to a quantum mechanical phenomenon known as the Pauli exclusion principle. This force is repulsive, like nuclear forces at very close range. The Pauli exclusion principle states that fermionic particles such as neutrons cannot occupy the same state, unlike the bosonic ones. As a result, the neutrons will only "pile up" on top of each other.

It has been realized since then that a neutron star is the end stage of a massive star after all its hydrogen is burnt and it has passed through the subsequent stages, burning helium, carbon, silicon, all the way to iron (see the discussion on supernova in Chapter 4). When the core reaches the stage where it contains only iron, nuclear burning can no longer provide energy to counteract gravitational attraction. The star at that stage continues to collapse under the force of gravity. In this process, as matter gets squeezed, most of the energy is released in the form of neutrinos and about 1% in the form of light. But even this small fraction of energy is so bright that it produces a spectacular firework display in the sky. This is called a supernova. After the supernova is over, depending on the mass of the original star, the remnant either becomes a neutron star or a black hole from which no light can escape.

Theoretically, therefore, there were strong reasons for a neutron star to exist. Its discovery, however, took time until it was finally found in 1967 by Jocelyn Bell and Anthony Hewish, who discovered a rapidly pulsing light from the sky. These objects were called pulsars, and it was eventually realized that the pulsars are nothing but neutron stars. They emit pulsing light because they rotate around their axes very rapidly. They also have an extremely intense magnetic field. More neutron stars have since been discovered: about 3,000 are now known. Some of them (very few) also come in pairs and are called binary pulsars [81].

A neutron star has one of the most exotic interiors of any astrophysical object. It is an extremely dense object: One cubic centimeter of it (roughly a teaspoon of neutron star matter) weighs about four billion tons. There is speculation that a neutron star is so dense that its core could consist of only quarks due to the melting of protons and neutrons into their constituents. Such a conclusion, however, remains quite controversial at the moment. Nevertheless, scientists agree that neutron stars are highly exotic objects, understanding whose interior could teach us a great deal about how matter behaves at high density.

Neutron stars are typically about 10–20 kilometers in size and have a mass of about 1.2–2 solar masses. Their mean densities are extremely high, about 100 trillion times that of water. They evolve as the universe evolves, slowly cooling, first by emission of neutrinos, followed by radiation cooling. Their surface temperatures vary depending on their age, but are, on average, about a million degrees Celsius compared to about 5,000 degrees for the surface of the Sun. The presence of a mirror universe has implications for neutron stars in our universe.

9.5.1. *Mirror Pollution in Neutron Stars*

Dark matter is present everywhere in the universe, and if it consists of mirror particles, it is stable. It is then quite possible that a dense object such as a neutron star or even a normal star will attract it due to its gravitational force. This can lead to its accumulation in neutron stars and create "impure neutron stars." This accumulation could occur in several ways: (i) a neutron star could simply attract mirror neutrons from the diffuse mirror neutron cloud surrounding it due to gravity, as noted; (ii) alternatively, if there is a mirror neutron object, say a mirror neutron star, its collision with a normal neutron star could lead to a violent explosion, leaving behind a mirror-polluted normal neutron star. There could be other mechanisms as well if there are some feeble forces between the neutron and the mirror neutron dark matter.

Such mixed stars will have two kinds of particles, each exerting gravitational forces on neutrons of its own kind as well as on the other kind, since gravity is a common force between the two kinds of neutrons. Each kind of neutron will also exert nuclear forces on neutrons of its own kind. Such systems were analyzed in Refs. [82, 83]. In Ref. [82], it was found that if the mirror dark matter mass is less than that of the familiar neutron, the total mass of the impure neutron star will be larger. A typical normal neutron star mass is about 1.4 times the mass of the Sun, which is known as the Chandrasekhar mass, and it could easily be between 1 and 2 solar masses. If neutron stars outside this range are discovered, that could be an indication of mirror dark matter. Note that if dark matter is either a supersymmetric particle and if it is captured in a neutron star, it could annihilate with another dark matter particle and emit energetic photons.

It turns out that there have been observations of heavier neutron stars. The neutron star PSR J0740+6620, which is about 2.1 solar masses, is one of them. Another one, the neutron star PSR J0952-0607, about 300 light-years away from the Earth and in the constellation Sextans, is about 2.35 solar masses. This may be the heaviest neutron star known to date.

Are the heavier neutron stars mixed neutron stars containing mirror matter? We do not know, and further study is needed. Similarly, studies show that if the mirror neutron mass is larger than that of the familiar neutron mass, the mixed neutron star mass could be smaller than 1.4 solar masses. One example of such a neutron star is HESSJ1731-347, with a mass of 0.77 solar masses.

Mirror pollution is also likely to lead to other kinds of observable effects. For example, their size might differ compared to pure neutron stars, and their surface temperature could also be affected. Planned instruments such as NICER (already launched since 2017), ATHENA, and STROBE-X (both planned to launch in 2031) are expected to measure the mass and radius of neutron stars with high accuracy. These observations will provide valuable insights into the properties of neutron stars. For example, by comparing the radii of two equal mass neutron stars, one may be able to conclude the presence of dark matter in one of them, offering interesting clues about the mirror world scenario. Also, astronomers have theoretical expectations for the mass–radius relation of pure neutron stars, and any deviation from these relations could signal the presence of dark matter within them. Thus, a detailed study of neutron stars could be a good way to test the mirror world hypothesis.

9.5.2. *Overheated Neutron Stars*

Now, let us discuss the implications for a neutron star in a symmetric mirror world scenario, where neutron–mirror neutron oscillation (n–n') can take place. In this case, we expect several exotic effects, for instance, some of the neutrons in the neutron star will convert to mirror neutrons. Once they become mirror neutrons, they will have no interaction with the surrounding neutrons and, under gravitational attraction, fall to the center of the neutron star. In this process, the mass of the neutron star will decrease. By Einstein's $E = mc^2$ formula, the mass loss will lead to the emission of energy from the star, and the star can heat up [84, 85].

This heating occurs due to the extra energy release. However, in symmetric mirror models, the mirror neutron can beta decay to mirror particles, and the resulting e'–e scattering can steal this energy from the neutron star and emit it as invisible dark mirror photons. This could lessen the heating effect. The magnitude of the heating will then depend on the rate of neutron–mirror neutron transformation, thus connecting an intrinsic particle physics phenomenon to astrophysics.

There is an International Space Station payload called Neutron Star Interior Composition Explorer Mission (NICER), which has been observing neutron stars since 2017 to measure their radii and temperatures to better understand the properties of these enigmatic astrophysical objects. It has already provided data on a neutron star called J0740 and furnished very valuable information about its interior. As a result of these observations, our understanding of neutron stars will improve, and this may eventually help us to explore different particle physics phenomena, such as n–n' oscillation and the possibility of a mirror universe.

9.5.3. *Slowing Down of a Rotating Binary Pair of Neutron Stars Due to n–n′ Oscillation*

Another specific effect of n–n' oscillation appears in a class of neutron stars that occur in pairs, called binary neutron stars. These stars have a certain orbital period as they go around each other. This can be determined from observations of their electromagnetic emission. It was pointed out by James Jeans in 1928 that their rate of rotation decreases and the orbital period increases if they lose part of their mass. This would happen if n–n' oscillation occurs, as shown in Fig. 9.3. As the mirror neutrons fall to the center, the mass of the neutron star decreases. This could also arise from other forms of energy loss. This effect can be observed since the orbital period of binary neutron stars can be measured. A famous example of such a measurement is that by Russell Hulse and Joseph Taylor, who not only discovered the first binary neutron star but also, through this discovery, provided the first indirect evidence for gravitational wave radiation, which caused the slowing down of their period. Hulse and Taylor were awarded the Nobel Prize in 1993 for this work.

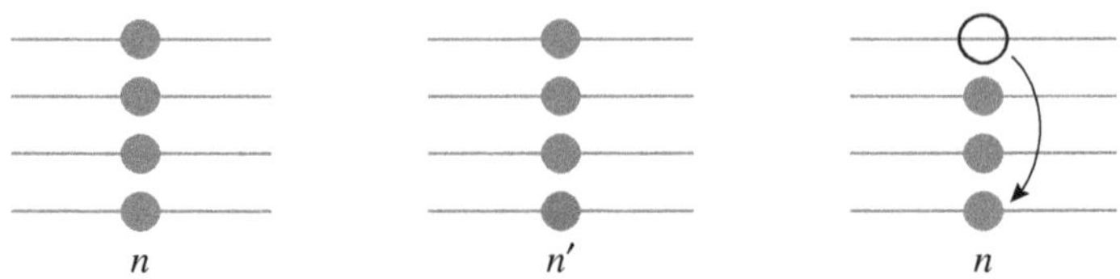

Fig. 9.3. Schematic illustration of n–n' oscillation inside a neutron star. The horizontal lines represent the energy levels within the star. The left panel shows the configuration before oscillation. The middle panel shows a neutron n converting into a mirror neutron n'. The n' then moves to a lower level, leaving a vacancy that is filled by a neutron from a higher level, releasing energy.

Currently, there are observational limits on how fast the orbital periods of observed binary neutron stars are slowing down, and from these, one can obtain limits on the rate of n–n' oscillation [86]. These limits are useful when experimentalists are looking for n–n' oscillation in the laboratory. For example, if they find that laboratory results are in disagreement with bounds from neutron stars, that would suggest an incomplete understanding of neutron stars. Such anomalies in various observations are how science progresses on various fronts.

9.6. Dancing with a Mirror Star

Stars do not always come alone, and it is believed that roughly one-third to one-half of the stars in a galaxy are binary stars, i.e., two stars orbiting each other. The name *binary star* was coined by William Herschel in 1802, and many more have since been identified. The important thing to know is that this is a purely gravitational effect, and since mirror stars can also gravitationally bind to normal stars, a new possibility arises: Some of the binaries could have mirror stars as partners of normal stars. In that case, only one partner of the binary will be visible, and the other will remain completely invisible. The presence of the partner, however, can be inferred by looking at the varying Doppler shift of light from the visible star as it orbits the mirror star. We will call them normal–mirror binaries. Since normal starlight can go through a mirror star, there will be no brightness variation due to the partner blocking the starlight. In fact, there may be a temporary brightness increase in one of the stars due to gravitational lensing if they are suitably located.

An example of a binary star with one star normal and its partner star being dark is the Epsilon Aurigae, which was first observed by the German astronomer Johann Frisch in 1821. He saw the variation in the brightness of the visible partner star of Epsilon Aurigae and suspected it to be a variable star. But later observations, especially those by Hans Ludendorff, revealed that it was a binary with the companion star being invisible and its eclipsing effect causing the variation in the visible partner's brightness. This binary has a period of 27 years, as deciphered from its brightness variations. It is not known whether the partner is a black hole or an invisible normal star. Until the nature of the partner star is confirmed, Epsilon Aurigae could be imagined as a potential candidate for a normal–mirror binary where the eclipsing effect may arise from a tiny photon–mirror photon mixing.

9.7. Sub-Solar Mass Black Holes

Black holes were described in Section 4.6 of Chapter 4 as candidates for dark matter. Two primary mechanisms for the formation of black holes were identified: (i) the death of stars after their complete evolution; (ii) mass density fluctuations during the early period of the universe's evolution. The latter are commonly referred to as *primordial black holes*. An important property of the black holes formed from the death of stars is that they possess a minimum mass of about one and a half (actually 1.4) times the solar mass, the aforementioned *Chandrasekhar mass*. In contrast, the primordial black holes are not subject to this constraint and can have masses below this limit. Recent gravitational wave observations have established that there are black holes with masses smaller than that of the Sun — hence the term sub-solar mass black holes [87]. This has prompted renewed interest in other possible origins of sub-solar black holes, and it turns out that this is relevant to the mirror baryons or mirror atoms as dark matter [88, 89].

If mirror particles such as mirror neutrons or mirror hydrogen atoms constitute dark matter, then they are very likely to belong to the class of asymmetric dark matter, as discussed in Chapter 4. This means that only particles act as dark matter, while their anti-particles have somehow disappeared, just like in the visible universe, where protons and neutrons persist while their anti-particles have largely vanished. The way sub-solar mass black holes can emerge in this scenario is as follows: Existing stars and other astrophysical objects can absorb the mirror dark matter owing to its feeble interaction with normal matter. As this dark matter accumulates inside the star, its density increases, and its gravitational force overwhelms any internal pressure that was holding the star prior to its capture, leading to the formation of a black hole. The resulting black hole mass depends on the mass of mirror dark matter particles and can therefore be smaller than the Chandrasekhar limit of 1.4 solar masses if these dark matter particles are heavier than protons. The discovery of more sub-solar mass black holes could provide evidence for the existence of a mirror world, offering valuable insights into the nature of dark matter.

9.8. Life in the Mirror Sector

Beyond cosmic diversity, the existence of a mirror universe could also have implications for the diversity of life. To explore this possibility, we must first define what we mean by life. As we look around on our planet Earth, we find that most life is based on carbon and its compounds. Our familiar form of life draws energy from the Sun to survive, sustain, and multiply. We cannot expect life everywhere to be of the same type. For instance, deep beneath the ocean floor, life has been discovered that does not rely on sunlight but instead derives energy from the Earth's interior. Thus, when we speak of life, we have to be careful about how we define it.

What, then, is a better definition of life? NASA defines life as follows: "Life is a self-sustaining chemical system capable of Darwinian evolution" [90, 91]. Notably, the ability to reproduce is not explicitly part of this definition. For example, viruses are often considered living entities, yet they are not self-reproducing and instead replicate only within the host cell of organisms.

With this minimal definition of life in mind, can we say anything about possible life in the mirror universe? We have argued that the mirror universe is likely to be very helium-rich, and consequently, all mirror stars would consist predominantly of helium. What would the existence of such a helium-rich universe imply for the existence of life? In particular, such a universe would likely have no water and therefore could not support carbon-based life familiar to us on Earth. However, this raises the question of whether other forms of life might still be possible.

Life does need a source of energy, which in the case of Earth is primarily supplied by sunlight. However, energy can be available in different forms as well, and this possibility must be considered when thinking about alternative life systems.

In a mirror universe composed largely of helium, any Sun-like star capable of providing life-sustaining energy would have disappeared after a relatively short time, perhaps about a million years. Could life in the mirror universe have disappeared along with such stars? Furthermore, could our understanding of life in the mirror universe inform us about the search for life on exoplanets in our own universe? As our understanding of the mirror universe improves, we may eventually find answers to these questions.

So, the answer to the question "Does the mirror universe have life?" is that we do not know, although the correct answer probably is that "it is unlikely".

Chapter 10

Mirror World and Death of Dinosaurs

> "The reality is you either step into the future, or you become a dinosaur."
>
> — **M. Shadows**

Could the mirror world idea have an impact on the paleontological history of the Earth? Paleontology is the study of the history of life on Earth based on fossils. An important question in paleontology has been: What caused the extinction of dinosaurs about 65 million years ago, as well as other periodic mass extinctions every 20 million years or so? There seems to be considerable evidence for them. In this chapter, we speculate that the mirror universe idea may provide an explanation for these events.

Dinosaurs are among the oldest types of vertebrates that once dominated the Earth many million years ago. Nobody is sure exactly when they first appeared on the surface of the Earth. The Earth is 4.5 billion years old, and some research suggests that dinosaurs may have first appeared around 200–230 million years ago. There were other vertebrates before them. In Australia, paleontologists discovered what they believe is the fossil of a 550-million-year-old frog. By that standard, dinosaurs are much younger and are believed to have appeared after the mass extinction of animals around 201 million years ago. There were about 1,000 different species of dinosaurs, most of which are non-avian (or not bird-like). Incidentally, birds are also believed to be a type of dinosaur; they are descendants of a group of flying dinosaurs called avian dinosaurs.

There are no more non-avian dinosaurs around in the world. In the museums, we only see their profiles reconstructed from fossils recovered

Fig. 10.1. Cartoon drawing of a non-avian Dinosaur, one of many that died during the mass extinction event around 65 million years ago.

from various places. From fossil dating, it has been established that they died *en masse* about 65 million years ago in some kind of catastrophic event that befell the Earth (Fig. 10.1). What kind of catastrophe it was, and how and why it happened, has puzzled scientists for many years, and rightly so, since it might teach us something about what could happen to us in the future. Parts of the answer to this question may be slowly emerging. One speculation discussed here is that it may be related to the existence of a mirror world.

Before we seek the origin of the catastrophic event, we need to get an idea of what kind of event could be so powerful that it would eliminate a whole group of species, from supersized to tiny vertebrates that once dominated the Earth. The story behind this reads like a detective novel.

Whatever caused the extinction of an entire group of animal species from the Earth's surface must have been a major event that dramatically changed the environment on Earth, making it difficult for these species to survive.

To change the environment in such a way, one possibility that has been speculated upon is that a huge amount of dust or other chemicals rose into the atmosphere and covered the Earth, blocking sunlight for a long time and drastically changing the climate. This could have been caused by an asteroid hitting the Earth.

Another possibility is that there was a large volcanic eruption that filled the atmosphere with toxic gases and caused many plants to die, killing plant-eating animals such as dinosaurs. There is evidence that such a major volcanic activity took place in northern India about 70 million years ago.

There have also been other theories based on widespread epidemic outbreaks or a supernova explosion in the sky. Certainly, in the wake of the deadly COVID-19 outbreak that killed millions of people, the epidemic theory does not seem so far-fetched.

The evidence, however, seems to point more toward the first kind of event than the others, and we discuss it in the following to see how it could be connected to the mirror world hypothesis.

The asteroid theory started with the father–son science collaboration team of Luis Alvarez and Walter Alvarez, who postulated in 1980 that the event that sealed the fate of dinosaurs was caused by a large heavenly body, such as an asteroid, hitting the Earth. It probably struck the Earth with such force that it raised a huge amount of dust at the point of impact. That dust could have blocked sunlight, filled the atmosphere with deadly gases, and vaporized rock and dust. This would have affected the climate on Earth so drastically that it killed large animals such as dinosaurs.

Wherever the asteroid collided with the Earth, it must have created a large crater on the Earth's surface. Within this crater, there might have been chemicals that were originally present in most asteroids. These could provide valuable clues for scientists to pursue in order to validate or disprove this hypothesis.

Luis Walter Alvarez was born in San Francisco in 1911 and was an experimental particle physicist by training. He was a Professor at the University of California, Berkeley. He was awarded the Nobel Prize in 1968 for his work in particle physics. He also participated in the Manhattan Project at Los Alamos during World War II. His son, Walter Alvarez, is a geologist born in 1940, who also served as a Professor at UC Berkeley.

The father–son team postulated the asteroid theory of dinosaur extinction in 1980; the same theory was also independently given by Jan Smit, a Dutch paleontologist, in the same year. Evidence supporting this idea has since appeared at different locations around the world. The catastrophic event is believed to have occurred during the Cretaceous–Paleogene period of the Earth, about 65 million years ago. An asteroid is supposed to have struck the Earth at Chicxulub in the Yucatan Peninsula of Mexico. In 2010, a team of scientists, after a long study of the Chicxulub site, endorsed this hypothesis. They determined that a 10–15 kilometer-wide space rock struck the Earth at Chicxulub with a speed of about 20,000 miles per hour, releasing enormous energy equivalent to 100 megatons of TNT that darkened the Earth and covered it with a cloud of dust. The power of this impact was about a billion times greater than the energy released by the atomic bombs dropped over Hiroshima and Nagasaki. In 2016, drilling into the crater caused by this collision, scientists found chemicals that are present in asteroids, providing further evidence for this theory.

An alternative hypothesis for mass extinction is the entirely Earth-based cause of volcanoes, as stated above, and is known in the field as volcanism. There have also been suggestions that a fall in the sea level could have contributed to this extinction. We probably do not know for sure which was the real cause, or whether it was a combination of many factors. However, the Alvarez theory of an external object striking the Earth appears to be supported by considerable evidence. As we will see in the following section, the Alvarez idea fits neatly into the mirror universe model.

10.1. Other Periodic Mass Extinctions

The dinosaur extinction 65 million years ago was apparently not an isolated event in the history of Earth. Earth is 4.5 billion years old, and studies of species evolution have shown that there were other extinction events as well. It appears that these events may have a periodic pattern, as was pointed out by David Raup and Jack Sepkowski in an article in *Proceedings of the National Academy of Science* in 1984. From the fossil records of marine vertebrates, invertebrates, and protozoans going back about 250 million years, they noted that there had been 12 major mass extinction events. Their studies also indicated that five of these were particularly large events. Later studies covering longer periods of Earth's history have suggested that there may have been a total of about 20 mass extinctions, including some of smaller magnitude as well. They estimated the interval between the events to be about 26 million years. This analysis was refined in 2010 by Mellot and Bambach, who extended the Raup–Sepkowski study to 500 million years and concluded that the average interval between two consecutive such events was about 27 million years, with higher statistical accuracy. This was further developed by Michael R. Rampino, Ken Caldeira, and Yuhong Zhu in 2020. They analyzed the non-marine tetrapod data (amphibians, reptiles, birds, and mammals) using sophisticated mathematical techniques and concluded that the rough periodicity of mass extinction is about 27.5 million years. As noted, all these studies generally agree that five of these extinctions were major events, including the one that wiped out the dinosaurs about 65 million years ago. This event is known as the Cretaceous–Paleogene extinction event. The remaining events were also significant, though not as "intense" as these five.

The most recent such extinction is believed to have taken place about 11 million years ago. That would place the Earth now roughly halfway between two such extinctions: the most recent one and the next one coming.

The accumulating evidence for periodic extinctions raises the question of what their origin could be. More specifically, why are these events periodic, and why is the period so long? A prevalent idea, inspired by the long time scale involved, is that these extinctions may have astronomical causes. This is quite a plausible hypothesis. In other words, something catastrophic is happening every 27.5 million years in our galactic environment. One possibility that has been suggested is that Sun and its planets cycle through the mid-plane of the Milky Way galaxy every 30 million years. The mid-plane is a crowded field containing many astronomical objects that could trigger massive comet or asteroid showers, potentially leading to extinction events. However, this idea remains controversial, since the interval between these crossings does not match the 27.5-million-year periodicity of extinctions. According to some astronomers, the last time the solar system crossed the galactic plane was about 3 million years ago, and it will not cross it again for another 30 million years. No mass extinction seems to have occurred 3 million years ago.

10.2. Where Could the Asteroid Have Come from?

The next question to ask is: If the event was caused by a massive asteroid hitting the Earth, where could the asteroid — or a 15-kilometer-wide heavenly body as described above — have come from? In space and within the solar system, there are many kinds of natural objects left over from the time of the formation of the solar system. The nine planets are the ones we hear about more, but there are other objects as well. Two of the most prominent extended systems of such objects are the Kuiper belt, named after its discoverer Gerard Kuiper, and the Oort cloud, named after Dutch astronomer Jan Hendrik Oort. The Kuiper belt is a disk-shaped object, full of small comets and dust, which is relatively stable against disturbances from nearby massive objects. The Oort cloud is a spherical, much more extended system of objects of various sizes, including comets and asteroids. It is believed to be the source of many long-period comets, unlike the Kuiper belt, which mostly hosts short-period comets. When another astronomical object passes by the Oort cloud, the objects in it can be dislodged from the cloud by gravitational forces and sent toward planets around the Sun, including the Earth. The asteroid responsible for the extinction of dinosaurs is believed to have originated from such a dislocation in the Oort cloud.

Dutch astronomer Jan Hendrik Oort was born in 1900 and died in 1992. He studied at the University of Groningen and later became a Professor at the University of Leiden. In addition to proposing the Oort cloud, he also confirmed the theory that the Milky Way galaxy rotates in its own plane, thus disproving that the Sun was at the center of the Milky Way galaxy. He also established that the Sun takes about 225 million years to orbit the center of the Milky Way. He contributed many other ideas to astronomy and was undoubtedly one of the most distinguished astronomers of the past century.

10.3. The Nemesis Star, Mirror World and Dinosaur Extinction

Soon after the Raup and Sepkowski paper, a hypothesis was advanced by astronomers Davis, Hut, and Muller in 1984, suggesting that the extinction events could be due to a red or brown dwarf star orbiting the Sun slightly beyond the Oort cloud with a period of 27.5 million years. When it comes near the Oort cloud, it could dislodge a huge number of comets, potentially causing mass extinctions.

A brown dwarf typically has a mass somewhere between a planet and a star, and is not easily visible in a normal telescope. It is faint and can appear red or orange. It could have a mass around 7–10% of a regular star and about 15–75 times the mass of a planet like Jupiter. It is not massive enough to sustain nuclear reactions at its center, so it does not shine brightly like a normal star. About 1,800 brown dwarfs have been identified using infrared telescopes.

This star believed to be responsible for mass extinctions is often called the Nemesis star. The brown dwarf hypothesis would seem to imply that the Sun and the Nemesis star form a binary pair. However, it turns out that there are relatively few such binary stars in the universe, making it look like a coincidence that the Sun is part of one. Besides, such brown dwarfs should be detectable in infrared surveys, yet the Infrared Astronomical Satellite (IRAS) launched by NASA, as well as other subsequent astronomical surveys, failed to discover any brown dwarfs in our neighborhood of the sky. Future astronomical surveys will continue the search for such a dwarf star.

The question that then arises is: If there is a Nemesis star and it is not a brown dwarf that evaded all the searches by infrared telescopes, then what could it be? The Nemesis star is supposed to be about 1.5 light-years away from the Sun, roughly 25,000–150,000 times the distance from the Sun to the Earth. Its mass is estimated to be about half or less of the solar mass. Despite searches by the IRAS satellite launched in 1984 or WISE, the Wide-field Infrared Survey Explorer satellite launched in 2009 by NASA, there has been no evidence of a brown dwarf matching the Nemesis star.

This is where the idea of the mirror universe comes to the rescue and provides a simple way out. If the Nemesis star were a mirror star, as suggested by Zurab Silagadze in his talk at the Gran Sasso Summer Institute, it would exert gravitational effects like all normal stars, but would evade detection in all infrared surveys because mirror objects cannot be seen with our light or our infrared telescopes. As discussed in a previous chapter, one possible way to ensure their existence would be through microlensing techniques, as discussed in Chapter 9. This is because the gravitational lensing of the source star depends only on the mass of the mirror star and not any of its other properties.

10.4. Implications for the Future of Life on Earth

If the periodicity of these extinction events is correct, we may expect the next such extinction by a comet strike in about 14 million years, which could represent another turning point for life on Earth. Note that there is another such catastrophic turning point that comes about due to hydrogen burning in the Sun. The Sun provides light and warmth because hydrogen in its core is converted into helium, releasing enormous amounts of energy. This energy not only prevents the gravitational collapse of the Sun but also provides warmth and light essential for life on Earth. In fact, the Sun converts around 600 million tons of hydrogen into helium every second. At this rate, all the hydrogen at the core of the Sun will be exhausted in about 5 billion years. Once the hydrogen burning is over, the steady state situation in the Sun will change. The Sun's core shrinks, and helium burning starts giving out more energy. This extra heat will begin to evaporate the water on Earth, the Earth's atmosphere will become thinner, and more of the Sun's rays will reach the Earth's surface. Earth will become less friendly to life. Eventually, the Sun will become a red giant star due to helium burning and expand its outer surface to Mars, engulfing the Earth. This will burn all life on Earth to ashes unless the Earth's life has by then migrated to other planets farther away. Another possibility is that the Sun may shed a part of itself, in which case, the Earth's orbit will move farther away, and depending on how far it moves, life may still survive.

The mass extinction caused by comet strikes, however, is likely to come in another 14 million years, which is much before the Sun finishes burning. While we may escape the expansion of the Sun by migrating to other planets, escaping the effects of Oort cloud comet strikes might be much more difficult, since such comets could strike many planets in the solar system.

Chapter 11

Can We Communicate with the Mirror Sector?

> "Half the world is composed of people who have something to say and can't, and the other half who have nothing to say and keep on saying it."
>
> — **Robert Frost**

An all-important final question in the mind of any reader of this book would be: If there is such a mirror world nearby (or even overlapping with ours), can we communicate with it, or even more ambitiously, travel to it? So far, we have seen that the only common force connecting visible and mirror particles is gravity. However, though this force is important for large objects such as stars and planets, it is far too weak to help us communicate at the microscopic level. We can, of course, observe gravitational waves produced by astrophysical events in the mirror universe and draw conclusions about its structure. Perhaps, at some future stage, gravitational waves may provide a way to communicate between the two worlds, but for that, we would first need a much deeper understanding of them.

This is true for a bare minimum version of the mirror universe model. However, since we know so little about a possible mirror world, could there be other possibilities for communication that are not yet ruled out by observations in the visible universe? Could there be other forces that connect the two universes? Two such possibilities discussed in the literature are photon–mirror photon mixing and neutrino–mirror neutrino mixing.

11.1. Photon–Mirror Photon Mixing

Mirror photon–normal photon mixing is a theoretically allowed possibility that would make the particles of the mirror world weakly visible.

This is because most of the mirror particles would then acquire what we earlier called a *milli-charge*, which is a minute fraction of the electron charge. Due to the milli-charge on mirror matter, the normal photon can bounce off the mirror matter, however weakly that may be, thereby making it visible ever so minimally. This can have observable consequences, since the mirror matter will not be invisible anymore. Instead, it will be very weakly visible, depending on how weak the photon–mirror photon mixing is.

In principle, one could even send radio signals to the mirror world, which would be perceived very weakly, depending on the strength of photon–mirror photon mixing. That would give us a way to communicate with the mirror world.

11.1.1. *Light Shining Through Walls*

The photon–dark photon mixing would imply, as already alluded to, that parts of our light beam can suddenly disappear into the mirror light (mirror photon) and later reappear as light again. If the light beam disappears, the photon goes into a mirror photon state. Since the mirror photon has no interaction with normal matter, it can pass through a wall placed in the path of the mirror photon beam and reappear on the other side of the wall as a normal photon. Thus, to us, it would appear as if light has passed through the wall and emerged on the other side. For the known photon, of course, this is impossible. An experiment designed to test this possibility is called a "light shining through the wall" experiment [92]. In this experiment, a beam of light will be incident on an opaque wall, and there will be a sensor on the other side of the wall looking for light to "shine through." If the mirror photon converts to a normal photon on the other side, we will see light [92, 93]. It is an amazingly simple concept and is very similar to the neutron "shining through the wall" idea discussed earlier. The experimental efforts for observing such a process are underway at the German laboratory DESY [94]. A schematic of the setup is shown in Fig. 11.1.

So far, no signal has been observed in this experiment. There are currently strong limits on the strength of this conversion rate derived from observations of Supernova 1987A, particularly if the mirror photon has a

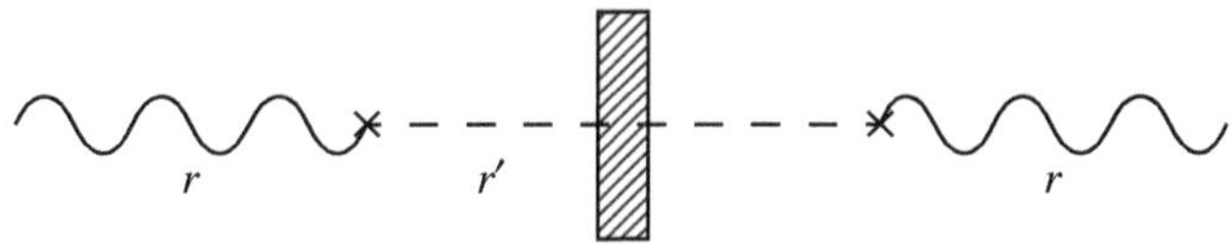

Fig. 11.1. A schematic of the "light shining through the wall" setup. The wavy line represents the familiar light wave, and the dashed line represents the mirror photon. The cross marks the location where a normal photon converts into a mirror photon.

mass greater than twice the electron mass. If those limits are considered, the probability of observing this process in the laboratory experiments is very slim. However, the mirror photon could be lighter than twice the electron mass, and therefore, a laboratory search is always important. Also, there could be unknown physical processes going on in the supernova environment that would invalidate the above-mentioned constraints. Thus, as detector sensitivities improve, it may eventually become possible to detect such mixings even at much lower strengths.

Mirror photon–normal photon mixing makes particles of the mirror world weakly visible, as noted earlier, and this could have a number of observable consequences.

If there exists a non-zero conversion between mirror photons and normal photons, other possible signatures of a mirror world could also be discovered. For example, consider a mirror asteroid passing through the Earth. Without the possibility of photon-mirror-photon mixing, such an asteroid would leave almost no trace apart from a weak gravitational tilt effect on the Earth, depending on its size. However, if mirror photons can convert into normal photons, the mirror asteroid could leave a hole in the ground due to the conversion of the mirror photons within the object into normal photons. Mirror matter can push aside normal matter through weak electric forces. To remind the reader, when we dig a hole in the ground, we do so by pushing a sharp object that displaces the atoms in the soil through electric forces. Similarly, a mirror asteroid can also "dig a tiny hole" in the ground as it passes through the Earth.

Other possibilities arising from photon−mirror photon mixing include mirror stars being visible like faint white dwarfs due to the milli-electric charge on the dark matter (e.g., mirror protons) [95]. The Gaia satellite is currently surveying the sky and collecting data. According to the European Space Agency, "Gaia is an ambitious mission to chart a three-dimensional map of our Galaxy, the Milky Way, in the process revealing the composition, formation and evolution of the Galaxy. Gaia will provide unprecedented positional and radial velocity measurements." These data can provide valuable information and offer clues about the possible existence of mirror stars.

If the photon–mirror photon mixing leading to light shining through the wall were ever discovered, it could also have many important and far-reaching practical applications. Depending on the strength of the mixing, it might even become possible to explore "objects behind a wall" in real life.

Photon–mirror photon mixing could also leave an imprint on the cosmic microwave background (CMB), which has been measured with great precision by several experiments. The reason is that the mirror universe would likely contain a cosmic mirror photon background, produced either as remnants of the Big Bang or as a result of mirror stellar evolution. Through photon mixing, some of these mirror photons could convert to normal photons, potentially distorting the CMB. Although the CMB has been studied extensively and no such distortion has yet been detected, future experiments with even higher precision could provide a rich source of information on the mirror universe.

11.2. Neutrino–Mirror Neutrino Mixing

Similar to photon–mirror photon mixing, there is also the possibility that the known neutrinos of the standard model could mix with the mirror neutrinos in a theoretically consistent way. In that case, normal neutrinos could oscillate into mirror neutrinos. This would provide another possible way to communicate with the mirror world.

Such mixing could also help explain the sterile neutrino puzzle alluded to earlier. An important point to note here is that if sterile neutrinos are confirmed to exist and it is shown that the mirror universe idea lies behind them, then this will favor the asymmetric mirror model rather than the symmetric one, since in the symmetric mirror model, the neutrino–mirror neutrino mixing will be maximal and will affect the discussion of Big Bang nucleosynthesis (BBN). For example, maximal mixing will imply the existence of three extra sterile neutrinos fully contributing to the energy density of the universe during the BBN epoch. However, observations do not allow such a possibility. In fact, they permit fewer than about 0.3 extra neutrinos to account for the observed primordial abundances of deuterium, helium, and lithium in the universe.

11.3. Wormhole Connections

It takes about five hours to fly from New York to Los Angeles. However, scientists have speculated that if wormholes existed in space connecting the two cities, then the journey could theoretically take only a matter of seconds. So, to understand this idea, we need to first understand what wormholes are.

Wormholes are connections between space–time points different from the obvious paths that we see in three dimensions. They are predicted by the wildly successful theory of Einstein and are sometimes also called Einstein–Rosen bridges after the 1935 work of Albert Einstein and Nathan Rosen, which first pointed out this possibility in the context of Einstein's theory of gravity. Wormholes provide shortcuts between space–time points or even shortcuts between universes in a multiverse scenario.

Einstein's theory of gravity endows space–time with a certain malleability like a rubber band. Space–time can bend and curve if there is a large object nearby, exerting huge forces. This lets the particles follow the shortest path on this curved space. So, instead of saying that the particle's path bends due to gravity, one can say that the particle moved in a curved space. This is an alternative way of looking at the particle being attracted by the force so that its path is bent. Thus, in Einstein's way of looking at things, force and space are interchangeable, in a manner of speaking. This idea of malleable space–time was so revolutionary that it gave rise to various new concepts and possibilities. One very famous example is the black hole. Black holes contain so much mass that the space around them becomes completely bent and warped. Light cannot come out of it because it becomes trapped and just keeps going round and round in the curved space around a black hole.

The concept of a wormhole emerged from the same line of reasoning. Einstein and Rosen found a simple solution of Einstein's theory that led to this idea in 1935. A wormhole can be visualized as a kind of a tunnel between two space–time points, regardless of the actual spatial distance of separation between the two points (e.g., New York and Los Angeles). The tunnel can have a very short length and be used as a shortcut between the two points. A schematic illustration is shown in Fig. 11.2. From this figure, one could visualize a possible way of connecting the two universes, A and B.

Einstein–Rosen wormhole connected a black hole with a white hole: A black hole attracts anything around it and does not let it go, whereas a white hole does not let anything enter its space, a sort of an opposite of the black hole, an anti-black hole. They argued that every black hole is accompanied by a white hole. This wormhole connecting two such objects was, however, shown to be unstable and to collapse in no time.

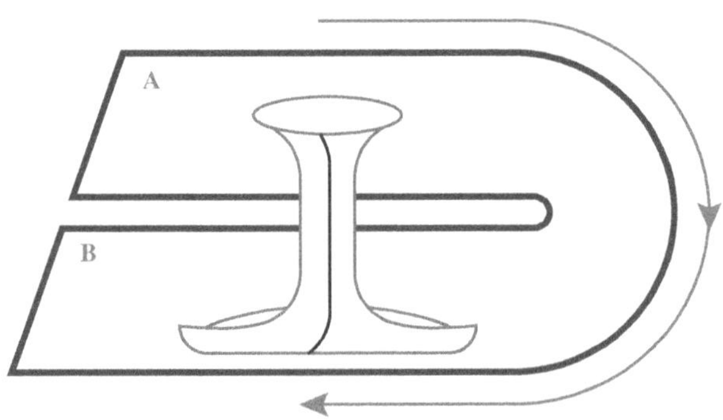

Fig. 11.2. A schematic illustration of a wormhole connecting two sides, labeled A and B, of the world. The curve outside shows how far the points on both sides are in normal space–time and the solid vertical line shows how it gets shortened when connected through a wormhole.

Since then, physicists have found ways to stabilize a wormhole using exotic matter, and its detailed implications are still under active investigation [96–100].

The possibility of communication between universes via wormholes has been a widely discussed topic in the gravity theory community since the famous paper by Einstein and Rosen. However, there appears to be no consensus at present on how complicated or feasible such communication via the imaginary wormholes might be. Of course, it is important to remember that, unlike black holes, predicted by Einstein's theory, there is currently no observational evidence for wormholes or white holes.

Some scientists have speculated that wormholes might have been produced during the early turbulent moments after the Big Bang and then disappeared almost immediately. If some of them survived, however, they could be around today. Just as black holes are detected by the gravitational waves emitted by them or energetic particle beams produced by their enormous forces, wormholes might also possess similar detectable properties that could be discovered one day.

Despite the tentative and speculative nature of these new ideas, we can imagine stable wormholes and discuss their implications for the mirror universe model, assuming that evidence for wormholes will eventually appear. Suppose that the mirror universe lies on one side of the bent space–time in Fig. 11.2 (say side A) and our universe lies on the other side (side B). In principle, the distance between any two points in the two worlds could be so immense that they can, for all practical purposes, be imagined as two universes. The wormhole can then provide a way to communicate between the two universes in a way faster than just going through space itself (see the curve vs. the vertical line).

Chapter 12

Epilogue

"The journey of a thousand miles begins with one step."

— **Lao Tzu**

"Sometimes it's more about the journey than the destination."

— **Jamal Crawford**

The possible existence of more worlds than the one we inhabit has always held a great deal of fascination for thinkers over the centuries, mostly in imagination. So has the concept of many universes. This possibility has started receiving serious scrutiny in recent years within many scientific contexts, such as string theory, quantum mechanics, and the origin of dark matter in the universe. After a brief discussion of historical, mythological, and more recent scientific interest in the concept of many universes, this book focuses on the possibility of a two-universe scenario that arises from speculations regarding symmetries in particle physics and explores the hypothesis that the second universe (called the replicate universe here) is the home of dark matter in the universe.

The possibility of a second universe coexisting with our own, discussed here, is exciting, not only from intellectual and scientific points of view but also possibly from a practical point of view in the sense of the long-term future of our universe. It takes the question "Are we alone in the universe?" to a different level and rephrases it as "Is our universe alone?" and analyzes its consequences. A positive observational answer to this question would surely be of groundbreaking importance. This makes this topic worth discussing and its possible signals worth investigating.

In earlier centuries, intellectual reasoning was the only tool used by thinkers in this line of inquiry, though it was used quite prolifically. The book starts with an overview of these investigations, emphasizing the hold they had on the imagination of many thinkers over many centuries. The last century, however, saw things change direction as thoughtful intellection started in this arena from multiple angles. This book is an attempt to summarize the insights gained from such scientific cerebration and focuses on its connection to the question of dark matter in the universe. It analyzes and elaborates on this question, outlining its many implications as well as tests. So, can we conclude after reading this book that we know the answer to the question "Is our universe alone"? The response, of course, is that right now, we do not know. But what this book presents are some of the concrete arguments that make it a sensible question to ask and make it plausible that the answer could at some point be yes, possibly in the far future.

The second universe scenario explored in Parts II and III is called the mirror universe (or mirror world), motivated by symmetries in particle physics. Thanks to all the activity in this area, it is now possible to envision a second universe as a science-based possibility and not just as a mythical fantasy. The mirror universe could be real as opposed to, for example, the many worlds possibility envisioned in quantum mechanics, which is hardly likely to be real.

Experiments are underway whose positive results could be interpreted as evidence for the mirror universe. In the end, it may or may not turn out to be a reality, but only further scientific exploration can tell. It is, therefore, encouraging and gratifying that scientists all over the world are now examining this question more seriously than before and have made it part of their future research agenda.

The main arguments for a mirror universe are two-fold: First, the observation of dark matter in the universe suggests that there must be a "home" for it. Supersymmetry at one point provided a leading framework that was explored extensively, however there seems to be no evidence for it so far. The axion is another compelling possibility that is under active experimental scrutiny right now, but again, there is no evidence for it yet. A third distinct possibility that stands out is the mirror universe hypothesis. Even though there is no concrete evidence for this possibility either, it helps explain why dark matter has eluded detection in many experiments and the suspicion — as indicated by some observations — that dark matter may have self-interactions, thereby broadening the scope of dark matter physics.

The attractive mirror universe as a venue for dark matter, as explored in this book and in the literature, rests on the theoretical argument that in a whole two-universe picture, mirror reflection is a good symmetry of the weak interactions and provides unity of the weak force with all other forces of nature. If such a universe exists, mirror symmetry would define a form of dark matter with specific properties. Although such dark matter is difficult to access using conventional particle detectors, there may be astrophysical as well as laboratory signatures for a lucky choice of parameter values. For example, if a sizable photon–mirror photon mixing exists, and/or if there is a sterile neutrino into which the known active neutrino can oscillate, we can get a glimpse of the mirror universe even in the laboratory.

The fact that within such a scheme, one could visualize an invisible mirror star, which forms a binary with the Sun and can explain periodic mass extinction on Earth, is an intriguing advantage of this scenario. This, of course, would require confirmation by actual evidence. One prediction of this idea is the existence of other mirror stars that could produce microlensing effects; exploration of such possibilities has just begun and needs to be pursued more extensively. There could also be opportunities in searching for binary systems in which only one star is invisible. One could use gravitational wave signals from such systems, which may help in the study of their properties.

The mirror universe could also be considered as a manifestation of an extra space dimension. The five-dimensional worlds could have two islands, each with its own 3 + 1 space–time, like ours, and separated by a distance in the fourth space dimension. It is much like two airplanes, one flying a mile above the other. The vertical height is analogous to the distance between the two islands along the extra fourth space dimension. When one plane tries to deliver fuel to the other plane, it connects a pipe (much like a wormhole would connect them) and supplies the other plane.

The idea of the mirror universe is no doubt speculative, although quite plausible in the opinion of this author. Yet, a better knowledge of the mirror universe and its structure could also have practical implications. Consider, for example, the possibility that we indeed live in a two-universe scenario and a large astrophysical mirror object comes closer to the solar system. It could then lead to gravitational perturbations of stars and galaxies in the observed universe while remaining invisible, and this could have practical implications for our future. This, for example, could include forecasting possible collisions of the solar system with mirror stars or the outward perturbation of planetary orbits. Clearly, then, better knowledge of the mirror universe could provide a way around such catastrophic possibilities.

The second universe idea discussed in this book has been a fertile ground for many creative discussions, and although speculative at present, as emphasized earlier, it is far from being totally implausible. Searches for its various implications, therefore, need to be taken seriously and pursued. Meanwhile, an interesting theoretical generalization of this idea to an arbitrary number of universes has been proposed to solve problems of the standard model of particle physics [101].

Finally, what if a dark matter particle is discovered using conventional detectors? Since we used the lack of evidence for dark matter in the laboratory searches as one motivation for a second universe, would the discovery of a dark matter particle in laboratory or underground searches, negate the replicate universe idea? Not necessarily, since dark matter could have multiple components, with mirror matter as a possibly subdominant component. Second, whether the lightest mirror particle is itself a dark matter particle or not, the particle physics symmetry argument for the mirror universe can stand on its own. The "dinosaur death" scenario could certainly remain as a possible argument in favor of this subdominant mirror sector. Even if the mirror particle is not the dark matter, mirror stars and mirror planets could still form if there exists a mirror universe of particles with properties similar to those in the visible universe. From this vantage point, all the experiments proposed to search for the mirror particles, such as neutron–mirror neutron oscillation or photon–mirror photon transition, should continue as long as possible, given how profound the mirror universe idea is.

Notes and References

[1] Parts of this discussion are taken from R. Zed, "What do scriptures say about the existence of other universes?" *Reno Gaz. J.* (3 January 2021).

[2] For other interesting articles on multiverse idea in history and religion, see Nautilus (January 2017).

[3] M. Sedacca, "Nautilus" (January 2017).

[4] T. Siegfried, *The Number of the Heavens*, Cambridge, MA: Harvard University Press (2019).

[5] A. C. Sparavigna, "Robert Grosseteste's thought on light and form of the world", *Int. J. Sci.* **3**(4), 54–62 (2014).

[6] Stanford Encyclopedia of Philosophy (2007).

[7] J. Gribbin, *Six Impossible Things*, MIT Press (2019).

[8] P. Byrne, "Many worlds of Hugh Everett", *Sci. Am.* (21 October 2008).

[9] N. G. Everett, Letter to Frank J. Tipler dated October 10, 1983. Archive, Box 1, Folder 9, https://space.mit.edu/home/tegmark/everett/sources.html#70.

[10] S. Osnaghi, F. Freitas and O. Freire Jr., "The origin of the Everettian Heresy", *Stud. Hist. Philos. Mod. Phys.* **40**(2), 97–123 (2009), cited in Wikipedia.

[11] S. Carroll, *Something Deeply Hidden*, New York: Dutton (2019).

[12] Interview of Oskar Klein by J. L. Heilbron and L. Rosenfeld on 1963 February 25, Niels Bohr Library & Archives, American Institute of Physics, College Park, MD, USA, www.aip.org/history-programs/niels-bohr-library/oral-histories/4709-3.

[13] N. Arkani-Hamed, S. Dimopoulos and G. Dvali, *Phys. Lett. B* **429**, 263 (1998), hep-ph/9803315 [hep-ph]; L. Randall and R. Sundrum, *Phys. Rev. Lett.* **83**, 3370–3373 (1999), hep-ph/9905221 [hep-ph].

[14] B. Carter, "Large number coincidences and the anthropic principle in cosmology", in *IAU Symposium 63: Confrontation of Cosmological Theories with Observational Data,* Vol. 63, Reidel, Dordrecht, pp. 291–298. DOI: 10.1017/S0074180900235638; Republished online by Cambridge University Press (7 February 2017).

[15] J. D. Barrow and F. J. Tipler, *The Anthropic Cosmological Principle* (1st edn.), Oxford: Oxford University Press (1986).

[16] L. Susskind, "The anthropic landscape of string theory", arXiv:hep-th/0302219.

[17] APS News, "This month in physics history", **15**(6) (May 2006).

[18] D. Nolte, "The fall and rise of Doppler effect", *Phys. Today* **73,** 30 (2020).
[19] For excellent reviews of what we know about dark matter today as well as its history, see G. Bertone and D. Hooper, *Rev. Mod. Phys.* **90**(4), 045002 (2018); S. Profumo, *Introduction to Particle Dark Matter*, Singapore: World Scientific (2019); M. Cirelli, A. Strumia and J. Zupan, e-Print: 2406.01705 [hep-ph]; S. Nussinov, arXiv:2510.25877.
[20] Nobel prize.org.
[21] E. Adelberger *et al., Class. Quantum Grav.* **29**, 184002 (2012).
[22] For a history of the discovery of the Higgs mechanism, see F. Close, *The Infinity Puzzle*, Basic Books, New York (2011).
[23] J. C. Pati and A. Salam, *Phys. Rev. D* **10**, 275 (1974); R. N. Mohapatra and J. C. Pati, *Phys. Rev. D* **11**(566), 2558 (1975); G. Senjanovic and R. N. Mohapatra, *Phys. Rev. D* **12**, 1502 (1975).
[24] P. Minkowski, *Phys. Lett. B* **67**, 421 (1977); R. N. Mohapatra and G. Senjanovic, *Phys. Rev. Lett.* **44**, 912 (1980); M. Gell-Mann, P. Ramond and R. Slansky, *Supergravity* (P. Van Niewenhuizen and D. Freedman (Eds.)), North Holland, Amsterdam (1980); T. Yanagida, *Workshop on Unified Theories and Baryon Number in the Universe* (A. Sawada and A. Sugamoto (Eds.)), KEK, Tsukuba (1979); S. L. Glashow, *NATO Sci. Ser. B* **61**, 687 (1980).
[25] For a discussion of the various aspects of the neutrino story, see R. N. Mohapatra, *The Neutrino Story: One Tiny Particle's Grand Role in the Cosmos*, Cham, Switzerland: Springer-Nature (2020).
[26] E. Bulbul *et al., AJ* **789**, 13 (2014).
[27] A. Boyarsky *et al., Phys. Rev. D* **113**, 251301 (2014).
[28] R. D. Peccei and H. Quinn, *Phys. Rev. Lett.* **38**, 1440–1443 (1977).
[29] S. Weinberg, *Phys. Rev. Lett.* **40**, 223–226 (1978).
[30] F. Wilczek, *Phys. Rev. Lett.* **40**, 279–282 (1978).
[31] P. Sikivie, *Phys. Rev. Lett.* **51**, 1415–1417 (1983).
[32] B. Carr, *Annu. Rev. Astron. Astrophys.* **32**, 531–590 (1994).
[33] For a review of Blackholes as dark matter, see B. Carr, K. Kohri, Y. Sendonda and J. Yokoyama, *Rept. Prog. Phys.* **84**, 116902 (2021).
[34] B. Holdom, *Phys. Lett. B* **166**, 196–198 (1986).
[35] D. Gabrowska, T. Melia and S. Rajendran, *Phys. Rev. D* **98**(11), 115020 (2018).
[36] A. Drlica-Wagner, C.-P. Weinstein and H.-B. Yu *et al.*, arxiv:2209.08215.
[37] Pauli's comment after the discovery of parity violation in 1956. Source: L. M. Lederman and D. Teresi, *The God Particle: If the Universe Is the Answer, What Is the Question?* Interlude C. Houghton Mifflin Company: Boston.
[38] L. Okun, I. Kobzarev and I. Pomeranchuk, *Sov. J. Nucl. Phys.* **3**(6), 837–841 (1966); *Yad. Fiz.* **3**, 1154–1160 (1966).

[39] R. N. Mohapatra and G. Senjanovic, *Phys. Rev. Lett.* **44**, 912 (1980).
[40] Early references on mirror world models: T. D. Lee and C. N. Yang, *Phys. Rev.* **104**(1), 254–258 (1956); I. Kobzarev, L. Okun and I. Pomeranchuk, *Sov. J. Nucl. Phys.* **3**, 837 (1966); M. Pavšič, *Int. J. Theor. Phys.* **9**(4), 229–244 (1974); S. I. Blinnikov and M. Y. Khlopov, *Sov. J. Nucl. Phys.* **36**, 472 (1982); R. Foot, H. Lew and R. Volkas, *Phys. Lett. B* **272**, 67–70 (1991) and *Mod. Phys. Lett. A* **7**, 2567–2574 (1992); Z. G. Berezhiani and R. N. Mohapatra, *Physi. Rev. D* **52**, 6607–6611 (1995); R. Foot and R. Volkas, *Phys. Rev. D* **52**, 6595–6606 (1995); Z. Chacko, H. S. Goh and R. Harnik, *Phys. Rev. Lett.* **96**, 231802 (2006).
[41] Z. Berezhiani, "Through the looking-glass: Alice's adventures in mirror world," in M. Shifman *et al.* (Eds.), *From Fields to Strings, Circumnavigating Theoretical Physics,* Vol. 3, pp. 2147–2195 [hep-ph/0508233]; Z. Berezhiani, D. Comelli and F. Villante, *Phys. Lett. B* **503**, 362 (2001).
[42] H. An, S. L. Chen, R. N. Mohapatra and Y. Zhang, *JHEP* **03**, 124 (2010).
[43] R. N. Mohapatra and V. L. Teplitz, *Astro. Phys. J.* **478**, 29–38 (1997); *Phys. Lett. B* **462**, 302–309 (1999).
[44] R. Foot, *Int. J. Mod. Phys. A* **29**, 1430013 (2014).
[45] Z. Chacko, D. Curtin, M. Geller and Y. Tsai, *JHEP* **11**, 198 (2021).
[46] D. Curtin and J. Setford, *JHEP* **03**, 041 (2020).
[47] E. Kolb and M. Turner, *The Early Universe*, Boca Raton, Florida: Taylor and Francis (1994).
[48] S. Weinberg, *First Three Minutes*, New York: Basic Books (1977).
[49] A. Guth, *Phys. Rev. D* **23**, 347–356 (1981).
[50] The basic idea of inflation was in several papers prior to Guth's work: see for example D. Kazanas, *Proceedings of the Seminar on Proton Stability* (D. Cline (Ed.)), University of Wisconsin Madison (1978); K. Sato, *Phys. Lett. B* **99**, 66 (1981).
[51] A. Linde, *Phys. Lett. B* **108**, 389 (1982); A. Albrecht and P. Steinhardt, *Phys. Rev. Lett.* **48**, 1220 (1982).
[52] A. H. Guth and S. Y. Pi, *Phys. Rev. Lett.* **49**, 1110–1113 (1982).
[53] J. Navarro, C. Frenk and S. White, *Astrophys. J.* **490**, 493–508 (1997).
[54] D. Spergel and P. Steinhardt, *Phys. Rev. Lett.* **84**, 3760 (2000).
[55] S. Tulin and H.-B. Yu, *Phys. Rept.* **730**, 1 (2018).
[56] Z. Berezhiani, A. Dolgov and R. N. Mohapatra, *Phys. Lett. B* **375**, 26–36 (1996); K. S. Babu and R. N. Mohapatra, *Symmetry* **14**, 4, 731 (2022).
[57] H. Hodges, *Phys. Rev. D* **47**, 456 (1993).
[58] E. Kolb, D. Seckel and M. Turner, *Nature* **314**, 415–419 (1985).

[59] Z. Berezhiani and L. Bento, *Phys. Rev. Lett.* **96**, 081801 (2006).
[60] R. N. Mohapatra and R. E. Marshak, *Phys. Rev. Lett.* **44**, 1316 (1980). This process was mentioned in the context of matter-anti-matter asymmetry in V. Kuzmin, *Pisma Zh. Eksp. Teor. Fiz.* **12**, 335 (1970).
[61] N. J. Ayres, Z. Berezhiani, R. Biondi, G. Bison, K. Bodek, V. Bondar, P. J. Chiu, M. Daum, R. T. Dinani and C. B. Doorenbos, *et al.*, arXiv:2111.02794 [physics.ins-det]; G. Ban, T. Lefort, O. Naviliat-Cuncic, K. Bodek, S. Kistryn, M. Kuzniak, J. Zejma, N. Khomutov, P. Knowles and M. Rebetez, *et al.*, *Phys. Rev. Lett.* **99**, 161603 (2007), arXiv:0705.2336 [nucl-ex]; Y. N. Pokotilovski, *Phys. Lett. B* **639**, 214–217 (2006), arXiv:nucl-ex/0601017 [nucl-ex]; A. P. Serebrov, E. B. Aleksandrov, N. A. Dovator, S. P. Dmitriev, A. K. Fomin, P. Geltenbort, A. G. Kharitonov, I. A. Krasnoschekova, M. S. Lasakov and A. N. Murashkin, *et al.*, *Phys. Lett. B* **663**, 181–185 (2008), arXiv:0706.3600 [nucl-ex]; C. Abel *et al.*, "nEDM", *Phys. Lett. B* **812**, 135993 (2021), arXiv:2009.11046 [hep-ph]; P. Mohanmurthy, A. R. Young, J. A. Winger and G. Zsigmond, arXiv:2201.04191 [nucl-ex]; L. J. Broussard, J. L. Barrow, L. DeBeer-Schmitt, T. Dennis, M. R. Fitzsimmons, M. J. Frost, C. E. Gilbert, F. M. Gonzalez, L. Heilbronn and E. B. Iverson, *Phys. Rev. Lett.* **128**(21), 212503 (2022), arXiv:2111.05543 [nucl-ex]; Y. Kamyshkov, J. Ternullo, L. Varriano and Z. Berezhiani, *Symmetry* **14**(2), 230 (2022), arXiv:2111.01791 [hep-ph].
[62] M. Hostert, D. McKeen, M. Pospelov and N. Raj, e-Print: 2201.02603[hep-ph].
[63] F. M. Gonzalez *et al.* (UCNτ Collaboration), "Improved neutron lifetime measurement with UCN", *Phys. Rev. Lett.* **127**, 162501 (2021) and references therein.
[64] A. T. Yue, M. S. Dewey, D. M. Gilliam, G. L. Greene, A. B. Laptev, J. S. Nico, W. M. Snow and F. E. Wietfeldt, "Improved determination of the neutron lifetime", *Phys. Rev. Lett.* **111**, 222501 (2013); G. L. Greene and P. Geltenbort, *Sci. Am.* **314**, 36 (2016) and references therein.
[65] More recently a beam experiment at JPARC facility has carried out a beam experiment which gives a result in agreement with the results of the bottle experiment. Y. Fuwa *et al.*, arxiv:2412.19519. The final verdict is yet to come in.
[66] B. Fornal and B. Grinstein, *Phys. Rev. Lett.* **120**(19), 191801 (2018); *Phys. Rev. Lett.* **124**(21), 219901 (2020) [Erratum].
[67] Z. Berezhiani, *Eur. Phys. J. C* **79**(6), 484 (2019).
[68] A. Bodas, M. A. Buen-Abad, A. Hook and R. Sundrum, *JHEP* **06**, 05 (2024).

[69] R. N. Mohapatra and N. Okada, *Phys. Rev. D* **111**(12), 123510 (2025).
[70] Asymmetric dark matter idea was suggested by S. Nussinov, *Phys. Lett. B* **165**, 55 (1985) and by S. Barr, R. Chivukula and E. Farhi, *Phys. Lett. B* **241**, 387–391 (1990).
[71] L. Wolfenstein, *Phys. Rev. D* **17**, 2369–2374 (1978); S. P. Mikheyev and A. Smirnov, *Sov. J. Nucl. Phys.* **42**, 913–917 (1985), *Yad. Fiz.* **42**, 1441–1448 (1985).
[72] C. Giunti and T. Lasserre, "eV-scale sterile neutrinos", arXiv:1901.08330 [hep-ph].
[73] K. N. Abazajian *et al.*, "Light sterile neutrinos: A white paper", arXiv:1204.5379 [hep-ph].
[74] Z. G. Berezhiani and R. N. Mohapatra, *Phys. Rev. D* **52**, 6607–6611 (1995).
[75] E. Baracchini *et al.*, "PTOLEMY expt" (2018), https://arxiv.org/pdf/1808.01892.pdf.
[76] C. Alcock *et al.*, *Ap. J.* **486**, 697 (1997); R. Ansari *et al.*, *A&A* **314**, 94 (1996).
[77] M. Mayor and D. Queloz, *Nature* **378**, 355 (1995).
[78] http://cfa-www.harvard.edu/planets/.
[79] A. Boss, *Science* **287**, 360 (1995).
[80] R. Foot, arXiv:astro-ph/0406257.
[81] University of Maryland website by Cole Miller, https://pages.astro.umd.edu/~mcmiller/nstar.html.
[82] I. Goldman, R. N. Mohapatra, S. Nussinov, D. Rosenbaum and V. Teplitz, *Phys. Lett. B* **725**, 200–207 (2013).
[83] P. Sandin and P. Ciarcellutti, *Astropart. Phys.* **32**, 278 (2009); *Phys. Lett. B* **695**, 19 (2011).
[84] D. Mckeen, M. Pospelov and N. Raj, *Phys. Rev. Lett.* **127**, 061805 (2021).
[85] I. Goldman, R. N. Mohapatra, S. Nussinov and Y. Zhang, *Phys. Rev. Lett.* **129**, 061103 (2022).
[86] I. Goldman, R. N. Mohapatra and S. Nussinov, *Phys. Rev. D* **100**, 123021 (2019).
[87] C. Kouvaris, P. Tinyakov and M. H. G. Tytgat, "Non-primordial solar mass black holes", *Phys. Rev. Lett.* **121**, 221102 (2018) [1804.06740].
[88] B. Dasgupta, A. Gupta and A. Ray, "Dark matter capture in celestial objects: Light mediators, self-interactions, and complementarity with direct detection", *JCAP* **10**, 023 (2020) [2006.10773].
[89] The LIGO Scientific Collaboration and The Virgo Collaboration, *Astrophys. J. Lett.* **892**, L3 (2020).
[90] S. A. Banner, "Defining life", *Astrobiology* **10**, 21 (2010).

[91] D. Clemente, arXiv:2301.05160 [astroph-EP].
[92] J. Redondo and A. Ringwald, *Contemp. Phys.* **52**, 211–236 (2011).
[93] J. Jaeckel and A. Ringwald, *Phys. Lett. B* **659**, 509–514 (2008).
[94] F. Januschek for the DESY Collaboration, arXiv:1410.1633.
[95] A. Howe, J. Setford, D. Curtin and C. Matzner, arXiv: 2112.05766.
[96] See for example, L. Susskind, arXiv:0503097 [gr-qc].
[97] J. Maldacena and A. Milekhin, *Phys. Rev. D* **103**(6), 066007 (2021).
[98] For a discussion of worm holes, see M. Visser, *Lorentzian Worm Holes*, New York: Springer-Verlag.
[99] See also the popular science book by M. Kaku, *Parallel Worlds*, Anchor Books, New York (2006).
[100] B. Greene, *The Fabric of the Cosmos: Space, Time, and the Texture of Reality*, Vintage Publisher, New York (2005).
[101] N. Arkani-Hamed, T. Cohen, R. T. D'Agnolo, A. Hook and H. D. Kim, *Phys. Rev. Lett.* **117**(25), 251801 (2016), arxiv:1607.06821 [hep-ph].

Glossary

Term	Definition
Anode	An electrode where positive electric charges reside.
Anthropic principle	The principle that states that the laws of physics are so designed as to be favorable for intelligent human life to exist and thrive.
Anti-matter	Matter composed of particles with the opposite electric charge; matter–antimatter pairs annihilate upon contact. Quantum theory and relativity dictate that every matter must have its anti-matter.
Asymmetric dark matter	The dark matter, whose anti-matter counterpart is absent, like the baryonic matter that the universe is made of.
Asymmetric mirror world	The mirror world scenario where particle masses differ from those in the visible sector.
Atom	The tiniest building blocks of all matter, made of protons, electrons, and neutrons.
Beta decay	Nuclear decay where one nucleus transforms to another accompanied by the emission of an electron and an anti-neutrino. Such decays are caused by the weak force in nature.
Big Bang	Beginning of the universe from an extremely hot and dense state.
Big Bang nucleosynthesis	The early-universe epoch, about one minute after the Big Bang, during which neutrons and protons combined to form heavier elements, such as helium, deuterium, lithium, and beryllium.

Binary stars	A pair of stars orbiting each other; Roughly half of the Sun-like stars are part of binary systems.
Black holes	A block of matter whose gravity is so strong that matter and light near it cannot escape its force of attraction, rendering them invisible.
Bosons	Particles with spin 0,1,2,... that act as carriers of forces.
Cathode	The electrode where the negative charges reside.
Cepheid variables	A class of pulsating stars whose pulsation periods are directly related to their brightness, making them useful for distance measurement.
Collapse of the wave function	The process by which a quantum system in a superposition of many different states appears in a single definite state upon measurement; not implied by the rules of quantum mechanics but added as an extra assumption.
Conservation	The principle that certain properties of a state, such as its total energy or total momentum of states involved, do not change in a scattering or any other physical process. This property is called a conservation law.
Cosmic microwave background	Remnant radiation from the early moments of the Big Bang that now pervades all space.
Dark energy	A form of energy that drives the current cosmic acceleration, believed to arise from a constant energy density that constitutes 70% of the total energy of the cosmos.
Dark matter	Hidden matter in the universe that leads to the anomalous motion of the stars in a galaxy and that of galaxies in a galactic cluster; constitutes 75% of the matter density in the current universe and provides the seed for galaxy formation.
Doppler shift	The frequency shift of the sound and light waves when either the source or the observer or both are moving; the frequency decreases when they move apart and increases when they approach.

Double beta decay	A nuclear process in which one nucleus decays into another by emitting two electrons. In standard double beta decay, each electron is supposed to be associated with an anti-neutrino; when the two electrons are not accompanied by two anti-neutrinos, it is called neutrinoless double beta decay, which, if observed, will provide evidence for lepton number violation and also for neutrinos being their own anti-particles. Intense search is on to discover neutrinoless double beta decay in various nuclei, such as germanium, molybdenum, and xenon.
Eigenstate	The state of a quantum system where it has a definite value for a particular observable or a particular set of observables.
Electrode	A conductor that allows electric current to enter a non-metallic medium, such as an electrolyte.
Exoplanets	Planets orbiting stars other than the Sun; more than 6,000 exoplanets have been discovered to date beyond the solar system.
Extinction of species	The disappearance of species from the Earth every 23 million years, as revealed by paleontological studies of mass extinction events over geological time.
Extra dimensions	Hypothetical additional spatial dimensions other than the familiar three; these can be real, but we do not see them because they are believed to be curled up to invisible sizes. Quantum theory predicts that the curled-up dimensions manifest as an infinite tower of massive particles called Kaluza–Klein particles.
Gauge invariance	A symmetry of a system which remains unchanged when independent transformations are carried out at each space–time point; this predicts the existence of a massless vector field. The carriers of weak and electromagnetic forces are known examples of forces mediated by these massless fields.

Gluons	The gauge boson that mediates nuclear forces. There are a total of eight gluons, which, like quarks, are not observed as free particles.
Grand unified theories	Theories that unify the weak, electromagnetic and strong interactions at very short distances. Many examples of such theories exist in the literature. At the moment, there is no experimental evidence for any grand unified theory.
Higgs boson	A spin-zero, electrically neutral particle responsible for giving mass to the various fermions and gauge fields in the standard model of electro-weak interactions.
Higgs spread	The constant value of the Higgs field, which is the same all over the universe and gives mass to the standard model particles and to any particle that interacts with the Higgs boson.
Hubble expansion	The expansion of the volume of the universe as time progresses. The rate of expansion is determined by the energy density in the universe at a given moment; as the expansion continues, the energy density goes down, and hence the expansion rate becomes slower. Hubble expansion creates space as it continues and is different from an explosion of a firecracker, which expands into already existing space.
Inflation	Exponential expansion of the universe just after the Big Bang, before the start of the Hubble expansion. Inflation makes the cosmic background radiation isotropic and is believed to have provided the seeds that eventually grew to become the galaxies in the universe. When inflation ends, the Hubble expansion is supposed to start.

Isospin	A symmetry of nuclear forces that manifests itself by giving nearly equal masses for proton and neutron, as well as the three π-mesons. Many other groups of particles with nearly the same mass have been discovered. These groups of particles with nearly equal mass are said to belong to iso-multiplets. The number of particles (N) in an iso-multiplet determines the isospin value I of the particle by the formula ($N = 2I + 1$). A slight difference in the masses between the members of an iso-multiplet arises due to electromagnetic forces, which do not respect isospin symmetry.
Isotopes	Atomic nuclei with the same number of protons and a different number of neutrons, e.g., deuterium, which has a neutron in the nucleus together with the proton, are isotopes of hydrogen whose nucleus has only a single proton.
Kaluza–Klein particles	An infinite tower of particles arising from a curled-up or compactified hidden extra space dimension.
Lepton	Fundamental particles, including electrons, muons, and tauons, as well as the three accompanying neutrinos. In beta decay processes, each charged lepton electron, muon, or tau lepton (e, μ, τ) is accompanied by the anti-particle of the corresponding neutrino (ν_e, ν_μ, ν_τ). Each charged lepton–neutrino pair is believed to have its conserved lepton number. Observations of neutrino oscillations have proved that individual lepton numbers are not conserved; only the total lepton number is. Theories beyond the standard model often predict that the total lepton number is not conserved. Whether the total lepton number is conserved or not is an important question, the answer to which is being sought in many different experiments.

M-dwarf stars — The smallest and coolest stars which burn hydrogen like the Sun but have much lower surface temperature. They have been frequent targets for exoplanet searches and are known to have on the average about 1.3 exoplanets per star.

Microlensing — A gravitational lensing effect that temporarily brightens up the source star to an observer on the opposite side of the object to the source star. Many microlensing events (probably more than 20,000) have been observed; this is used to detect exoplanets and has resulted in the discovery of more than 200 of them.

Nemesis star — A hypothetical companion star of the Sun, assumed to be orbiting it once every 23 million years. This is believed to have kicked out meteorites as it passes by the Oort cloud, thereby causing the extinction of species on Earth.

Neutralino — The super-partners of the Z-boson and Higgs boson; the lightest neutralino is a stable fermion and has all the properties required to be the dark matter of the universe.

Neutrino — The lightest neutral fermions in the standard model. There are three of them, (ν_e, ν_μ, ν_τ) associated with the three charged leptons (e, μ, τ). It was thought for a long time that they were massless, but a tiny mass for them was discovered in 1998 from the observation of their oscillations from one species to another.

Neutron star — A star made mostly out of neutrons, with a radius of 10 kilometers and a surface temperature of about 10 million degrees Kelvin. Their existence was suggested by Baade and Zwicky, and they were discovered in the mid-1960s. It was subsequently understood that neutron stars are the end stage of a star; they initially cool by neutrino emission, but at a later stage only by photon

	emission. Neutron stars rotate very rapidly and contain very large magnetic fields.
Parity	The transformation of a system under mirror reflection. Parity is a good symmetry of electromagnetic, nuclear, and gravitational forces, but not of the weak forces.
Positron	The anti-particle of the electron. When an electron and a positron come together, they annihilate each other, giving two photons.
Pulsar	A pulsating star discovered in the mid-1960s. They were proven to be neutron stars whose rapid rotation gives rise to pulsating light.
Quantum chromodynamics	The theory of strong forces that uses an unbroken local symmetry operating on quark color degrees of freedom.
Quarks	The basic constituents of protons, neutrons, and π-mesons. They are accompanied by their anti-particles, the antiquarks. The protons and neutrons have three quarks inside them, and mesons such as π- and K-mesons are made of a quark and an anti-quark. The quarks are not supposed to appear as free states but always remain confined inside nucleons and mesons.
Self-interacting dark matter	Those dark matter theories where the dark matter particles exert forces on each other. This cures the core–cusp problem as well as the too many satellites problem of non-interacting dark matter theories.
Spin	An internal degree of freedom, which is like a rotation of the particle, although it is not a real rotation. Its value comes in discrete units 0, ½, 1, 3/2, etc. The particles carrying integral spins, e.g., 0,1, 2,…, are called bosons, while those carrying half integral spins, e.g., ½, 3/2, etc., are called fermions.

Spontaneous symmetry breaking	A phenomenon in which the physical laws governing a system have a symmetry that is not shared by its lowest energy state. For example, a standing rod that has rotational symmetry: When the rod falls to the ground (its lowest energy state), it breaks the symmetry spontaneously. Another example is a circular dining table: When all the wine glasses are to the right of the diners, it has a rotational symmetry, but as soon as one person picks up the glass, the symmetry is spontaneously broken.
String theory	A theoretical framework, developed between the 1960s and the 1970s, in which fundamental particles are one-dimensional, tiny, wiggly string-like objects rather than point particles. So far, there is no experimental evidence in favor of this theory.
Supernovae	End of a star's life after its nuclear core burning reaches the iron nucleus. Typically, burning from the birth of a star converts hydrogen to helium, helium to carbon, carbon to silicon, and silicon to nickel and iron.
Symmetry	The property of a system that remains unchanged when looked at from different perspectives. What the symmetry is depends on the angles from which the observer looks at it.
Symmetric mirror world	The version of the mirror world scenario in which the mirror partners of all known particles have the same masses as their standard model counterparts.
Type Ia Supernova	An explosion of a white dwarf star in a binary star system after exceeding 1.4 solar masses by accreting matter, leading to a thermonuclear reaction.

W-boson	The gauge boson that mediates electric charge-changing weak forces. It takes a neutron to a proton and an electron to a neutrino.
Wave function	A mathematical description of the quantum state of a physical system. It is a function of space and time and any other property, such as spin.
Wormholes	Hypothetical passages in space–time connecting two worlds as predicted by Einstein's general theory of relativity. These are mathematical entities whose existence has not yet been established.
Z-boson	The gauge boson that mediates neutral weak forces. It keeps the proton, neutron, electron, and neutrino unchanged during weak interaction.

Index

About the Author

Rabindra N. Mohapatra is a Distinguished University Professor Emeritus at the University of Maryland, College Park, MD, USA. Mohapatra received his PhD from the University of Rochester, NY, in 1969 and completed his post-doctoral research at Stony Brook University, NY, and the University of Maryland, College Park. He was a Professor at the City University of New York before joining the University of Maryland in 1983. He became a Distinguished University Professor in 2016.

Mohapatra has worked extensively on neutrino physics and dark matter. He is one of the proponents of the seesaw mechanism for neutrino masses and the left–right symmetric theories of weak interactions. He is also the co-proponent of the new particle related to neutrino mass called the "Majoron." He suggested the idea of neutron–anti-neutron oscillation, which led to many experimental searches for this process. His interest in the mirror universe model started more than 30 years ago, and he has made original contributions to its implications for astronomy and particle physics.

Mohapatra is a fellow of the American Physical Society and the National Academy of Sciences of India (NASI). He was awarded the Humboldt Prize in 2005 for his research on neutrinos and the Distinguished Scientist Award from the American chapter of the Indian Physics Association in 2000. In 2008, he was awarded the degree of Doctor of Science (*honoris causa*) by North Orissa University, India.

Mohapatra is the author of the book *Unification and Supersymmetry: The Frontiers of Quark-Lepton Physics* (Springer-Verlag), whose third edition was published in 2003, and co-author (with Palash B. Pal) of the book *Massive Neutrinos in Physics and Astrophysics* (World Scientific), whose third edition was published in 2004. His most recent book is *The Neutrino Story: One Tiny Particle's Grand Role in the Cosmos*, published by Springer Nature (2020).

www.ingramcontent.com/pod-product-compliance
Ingram Content Group UK Ltd.
Pitfield, Milton Keynes, MK11 3LW, UK
UKHW061932100926
8556IPUK00007B/888

9 789819 832545